Lecture Notes in Mechanical Engineering

Lecture Notes in Mechanical Engineering (LNME) publishes the latest developments in Mechanical Engineering—quickly, informally and with high quality. Original research reported in proceedings and post-proceedings represents the core of LNME. Volumes published in LNME embrace all aspects, subfields and new challenges of mechanical engineering. Topics in the series include:

- Engineering Design
- Machinery and Machine Elements
- Mechanical Structures and Stress Analysis
- Automotive Engineering
- Engine Technology
- Aerospace Technology and Astronautics
- Nanotechnology and Microengineering
- Control, Robotics, Mechatronics
- MEMS
- Theoretical and Applied Mechanics
- Dynamical Systems, Control
- Fluid Mechanics
- Engineering Thermodynamics, Heat and Mass Transfer
- Manufacturing
- Precision Engineering, Instrumentation, Measurement
- Materials Engineering
- Tribology and Surface Technology

To submit a proposal or request further information, please contact the Springer Editor of your location:

China: Ms. Ella Zhang at ella.zhang@springer.com
India: Priya Vyas at priya.vyas@springer.com
Rest of Asia, Australia, New Zealand: Swati Meherishi at swati.meherishi@springer.com
All other countries: Dr. Leontina Di Cecco at Leontina.dicecco@springer.com

To submit a proposal for a monograph, please check our Springer Tracts in Mechanical Engineering at http://www.springer.com/series/11693 or contact Leontina.dicecco@springer.com

Indexed by SCOPUS. All books published in the series are submitted for consideration in Web of Science.

More information about this series at http://www.springer.com/series/11236

Avanish Kumar Dubey · Anish Sachdeva ·
Munish Mehta
Editors

Recent Trends in Industrial and Production Engineering

Select Proceedings of ICAST 2020

Springer

Editors
Avanish Kumar Dubey
Department of Mechanical Engineering
MNNIT Allahabad
Prayagraj, India

Anish Sachdeva
Department of Industrial and Production Engineering
Dr. B. R. Ambedkar National Institute of Technology Jalandhar
Jalandhar, Punjab, India

Munish Mehta
School of Mechanical Engineering
Lovely Professional University
Phagwara, India

ISSN 2195-4356 ISSN 2195-4364 (electronic)
Lecture Notes in Mechanical Engineering
ISBN 978-981-16-3134-4 ISBN 978-981-16-3135-1 (eBook)
https://doi.org/10.1007/978-981-16-3135-1

This Springer imprint is published by the registered company Springer Nature Singapore Pte Ltd.
The registered company address is: 152 Beach Road, #21-01/04 Gateway East, Singapore 189721, Singapore

Preface

With growing concerns about climate change and environmental degradation, sustainability has become a strategic priority for organizations around the world. Governments, consumers, and investors are now pushing automotive organizations to change their ways of working, culture, and products. So, where does the industry stand today in terms of its sustainability initiatives and what does the future hold?

International Conference on Advances in Sustainable Technologies (ICAST 2020) organized by Lovely Professional University, Jalandhar, Punjab, India, from November 6 to 7, 2020, aimed to answer this question and provide a leading forum to the scientists, researchers, academicians, industry professionals, and students across the globe for sharing original research contributions and practical developments in the field of mechanical engineering and sustainable technologies. The conference also provided an opportunity for national and international experts and industry leaders to share their experiences and success stories. ICAST 2020 played a key role to set up a bridge between academia and industry.

Due to the prevailing COVID-19 pandemic, the conference was organized virtually and hosted around 500 participants from India and across the world to exchange their scientific ideas. During 2 days of the conference, researchers from academia and industries went through various scientific brainstorming technical sessions and presented their most recent cutting-edge discoveries. This conference also provided a scope to establish a network for joint collaboration between academia and industry.

This book brings together the collection of cutting-edge research articles presented in ICAST 2020. This book, which caters to the industrial and production engineering aspects, will serve as a reference guide for researchers and practitioners and is expected to foster better communication and closer cooperation between academia and industry partners. It encompasses industrial and production engineering areas such as sustainable manufacturing systems, additive manufacturing, decision sciences, logistics and supply chain management, composites, micro-/nanomaterials, operations management, simulation, computer-aided engineering, rapid prototyping, manufacturing management and automation, metrology, manufacturing process optimization, machining and machine tools, casting, welding, and forming.

Allahabad, India Prof. (Dr.) Avanish Kumar Dubey
Jalandhar, India Prof. (Dr.) Anish Sachdeva
Jalandhar, India Dr. Munish Mehta

Acknowledgements

We would like to acknowledge all the participants who have contributed to this volume. We thank all faculty and staff of the School of Mechanical Engineering, Lovely Professional University, who have directly or indirectly helped us to accomplish this goal. We would like to express our gratitude to respected Chancellor Sh. Ashok Mittal sir and worthy Pro-Chancellor Smt. Rashmi Mittal mam for extending their generous support as without their blessings, it would not have been possible to complete this task. We sincerely thank honorable Vice-Chancellor Prof. (Dr.) Ramesh Kanwar and respected Executive Dean Dr. Lovi Raj Gupta for their wonderful suggestions at all stages. We also thank Dean and Head of School Dr. Vijay Kumar Singh and Deputy Dean Dr. Ankur Bahl for guiding us at all stages.

Last but not least, we thank the staff at Springer, in particular Ms. Rini Christy, for their help and support.

Though due care has been taken, still, there is always space for improvement. We would appreciate any suggestions from the readers for further improvement of this book.

Jalandhar, India — Dr. Munish Mehta
Jalandhar, India — Prof. (Dr.) Anish Sachdeva
Allahabad, India — Prof. (Dr.) Avanish Kumar Dubey

Contents

About the Editors

Dr. Avanish Kumar Dubey is currently a Professor at the Department of Engineering/Mechanical Engineering, Motilal Nehru National Institute of Technology. He has authored and co-authored multiple peer-reviewed scientific papers and presented works at many national and international conferences. His contributions have acclaimed recognition from honorable subject experts around the world. He is also actively associated with different societies and academies. His academic career is decorated with several reputed awards and funding. His research interests include production (manufacturing), advanced machining processes, laser materials processing, design of experiments, micro-manufacturing processes. He has authored nearly 100 research articles, chaired numerous technical sessions and delivered many lectures.

Dr. Anish Sachdeva is a Professor in the Industrial and Production Engineering Department at Dr. B. R. Ambedkar National Institute of Technology Jalandhar, India. His areas of interest are reliability and maintenance engineering, supply chain management, optimization and simulation of production systems, and quality management. He obtained his B.Tech. from Dr. B. R. Ambedkar National Institute of Technology (erstwhile REC Jalandhar), M.Tech. from Guru Nanak Dev Engineering College Ludhiana and Ph.D. from Indian Institute of Technology Roorkee (India). He has guided 65 M.Tech. and 14 Ph.D. students. He has over 150 publications in international and national journals and proceedings of international conferences to his credit. He has organized five international conferences at NIT Jalandhar as organizing secretary and around 20 short-term courses in his area of expertise. He has organized several workshops and training programs for academic institutes and companies. He has edited a number of journal papers of repute as guest editors as well as volumes of books with reputed publishers. He also shares responsibilities of editorial board member in various international journals.

Dr. Munish Mehta is presently working as an Associate Professor in the School of Mechanical Engineering, Lovely Professional University, Phagwara, Punjab. He completed his Ph.D. from I.K. Gujral Punjab Technical University in 2018 and M.Tech. from National Institute of Technical Teachers Training and Research

(NITTTR) Chandigarh in 2010. He obtained his B.E. from Nagpur University in 1997. His research interests include reliability engineering, optimization, stochastic modeling, mechatronics and genetic algorithm etc. A lifetime member of ISTE, he has published various research papers and book chapters in different national/international journals and conferences of repute. His academic achievements include serving as reviewer of many reputed journals. He has supervised and guided various dissertations and projects at PG and UG level. He has organized a national conference as organizing secretary. He has 7 years of industrial and 16 years of teaching experience to his credit.

A Comparative Analysis of Different Techniques of Thermal Barrier Coating

Priyanka Sharma, Vijay Kumar Dwivedi, and Dipak Kumar

Abstract Numerous methods are used to put the ceramic material upon the substrate to form a film of the insulating material over it which is named as the thermal barrier coating to increase the life of the component and to reduce the pollution if applied in engine components. The coating is done to fulfill certain flaws of the substrate and to provide the surface that has the tendency to take the high temperature during operation. Also, thermal barrier coating is used to reach the higher efficiencies for a particular device like engine. The objective to enhance the efficiencies is to provide a substitute that can decrease the use of conventional resources like, fossil fuels. In this study, distinct techniques for the deposition of thermal barrier coating have been compared on the basis of their characteristics and forte. The methods are presented with their properties in this paper. It is not necessary that same method is used for the deposition of the top coat as well as the bond coat but there can be two different methods based on the requirement of the coating. The techniques used for the deposition not only affect the life of the thermal barrier coating but also the life of the component. So, the merits and demerits are compared for the different techniques in the present study.

Keywords Thermal barrier coating · Coating techniques · Air plasma spray · Pollution

Nomenclature

APS	Atmospheric plasma spray
EB-PVD	Electron beam-physical vapor deposition
HVOF	High velocity oxygen fuel

P. Sharma (✉) · V. K. Dwivedi
GLA University, Mathura, Uttar Pradesh 281406, India

D. Kumar
RKGIT, Ghaziabad, Uttar Pradesh 201003, India

A. K. Dubey et al. (eds.), *Recent Trends in Industrial and Production Engineering*, Lecture Notes in Mechanical Engineering,
https://doi.org/10.1007/978-981-16-3135-1_1

KSC	Kinetic spray coating
TBC	Thermal barrier coating
TGO	Thermally grown oxide
YSZ	Yttria stabilized zirconia

1 Introduction

The heat in the engine or turbine is need to be trapped within the operating components to fulfill the requirement and for getting highest efficiency. The barrier that is provided to the heat flow in the form of a film of insulation over the substrate material is termed as thermal barrier coatings [1, 2]. The substrate is the base material of the object or the component of a particular mechanical device which is normally metallic in nature. Thermal barrier coatings are mainly applied to the parts of a heat engine and gas turbine to maintain the operating temperature. The purpose of application of thermal barrier coating is exhaust heat management [3]. The structure of a typical thermal barrier coating is shown in Fig. 1.

The main properties of thermal barrier coatings cover the following points, low thermal conductivity, low thermal diffusivity, strong adherence to the base metal, phase and chemical stability, thermal shock resistance during thermal cycling and protection from oxidation and corrosion. Due to the low thermal conductivity and high thermal expansion coefficient, the most amply material used for the thermal barrier coating is Yttria stabilized zirconia (YSZ) which is ceramic in nature and is used as top layer of the coating [4–7]. YSZ embodies 7–8% of Y_2O_3 by weight and the thickness of top coat is approximately 125–500 μm. But the layer between top coat and substrate is known as the bond coat, and this is metallic in nature that has thickness 50–125 μm [8].

Another main point of consideration is the method of deposition of thermal barrier coating. The points of consideration for the deposition methods have been provided in the following text [3].

- Particle velocity
- Powder size

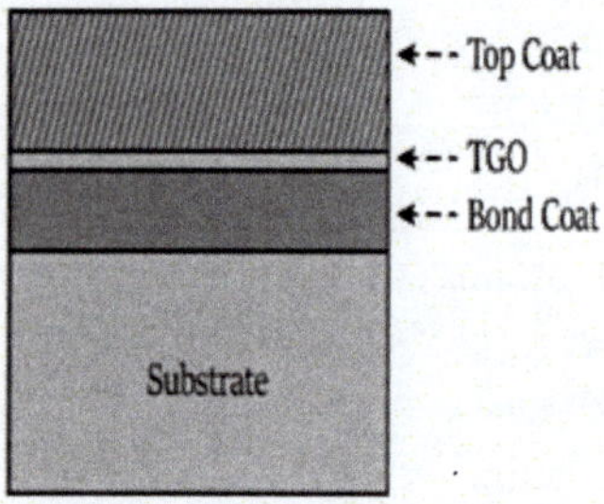

Fig. 1 Arrangement of layers in TBC [1]

- Process and feedstock
- Flame temperature.

To get the more bond strength and better adhesion, the particle velocity should be of higher range and the feedstock, and process plays an important role in the better bond strength and thickness of the coating. To get the incremented value of the hardness and wear resistance of the coating, there must be the lower range of the powder size of the coating material. Now, for the appearance of the coating, flame temperature should be high which is important for the fine appearance of the top coat.

The microstructure of the coating influenced by the following parameters [9],

- Torch current
- Plasma gas flow rate
- Carrier gas flow rate
- Torch transverse velocity
- Stand-off distance
- Material size
- Material temperature
- Material velocity.

2 Various Techniques of Coating

Numerous techniques for coating are being used for deposition of the top coat and bond coat on the metallic substrate. Thermal spray coating is most extensively used method for obtaining the thermal barrier coatings, and it is broadly classified as follows,

- Atmospheric plasma spraying techniques (APS)—Used for dense coating and it is the most extensively used method for the deposition of top layer in thermal barrier coatings.
- High velocity oxygen fuel (HVOF)—Used to get high wear resistance, corrosion resistance,, and erosion resistance, and it is most extensively used method for the deposition of bond coat (middle layer) in thermal barrier coatings.

On the basis of source of heat, the classification of thermal spray is given in flow chart in Fig. 2.

Also, some techniques are based on the diffusion which is also termed as pack cementation coating and it includes,

- Liquid state diffusion—predominantly used to obtain the coating in case of zinc, chromium, and copper substrate.
- Chemical vapor diffusion (deposition)—Used to obtain the thin layer over the surface of the substrate and the substrate for this type of technique is semi-conductor.

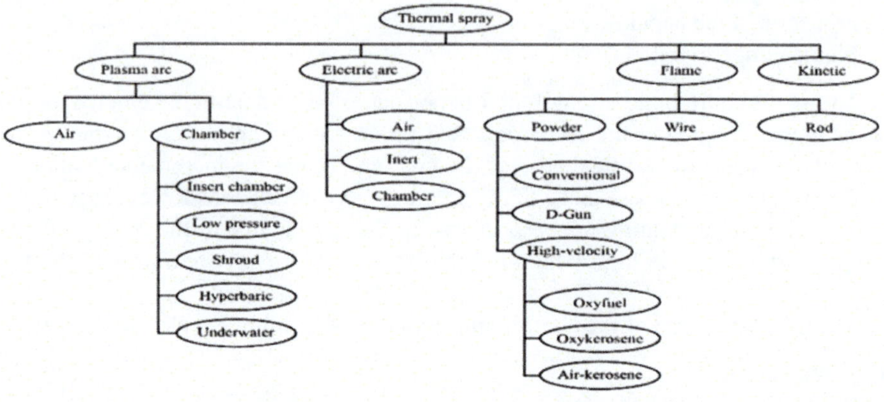

Fig. 2 Flow chart representing the types of thermal spray techniques [10]

Now, some techniques are also used, and they are, physical vapor deposition (PVD), magnetron sputtering coating, nickel dispersion coating, and electric arc wire spray coating.

2.1 *Atmospheric Plasma Spraying Technique (APS)*

The most cosmopolitan techniques to obtain the insulation coating that is thermal barrier coating over the metallic substrate are the conventional atmospheric plasma spraying technique [6, 7, 10–15]. It is one of the methods of thermal spray techniques that includes the spraying via plasma. In this techniques, inert gas extensively argon is used to generate the plasma by super heating the inert gas through DC arc [9, 10]. Also, a combination of argon and hydrogen gas is used to create plasma. The ceramic powder is sprayed over the substrate with help of a commercial gun. To operate the gun, 20–200 KW of the power is required [16].

The ionization of the inert gas is done between the anode and cathode through a high voltage that is pressurized through a particular nozzle, and then from the nozzle, it regains its previous state with high amount of heat energy [17]. At this time, the ceramic material for the coating is supplied to the arc in the form of powder so that arc can be able to melt the particles and project it well over the surface of substrate and consecutive splats of ceramic material are obtained on the metallic substrate only in one go, and hence, thickness can be increased by incrementing the number of pass [3]. The schematic diagram for the atmospheric plasma spraying coating as shown in Fig. 3.

The material of anode and cathode is copper and tungsten, respectively. Also, the argon gas can be replaced by nitrogen and helium depending upon the requirements of the coating. As the arc is having very high energy of heat so it can melt high refractory metal like tungsten and ceramic material like zirconia. The plasma coating when done

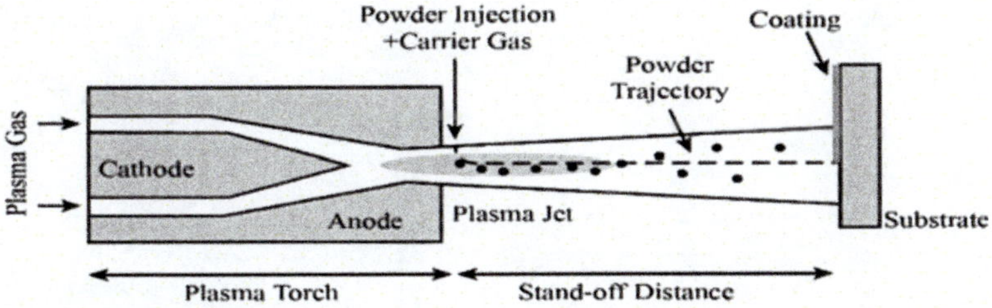

Fig. 3 Schematic diagram of atmospheric plasma sprayed coating [17]

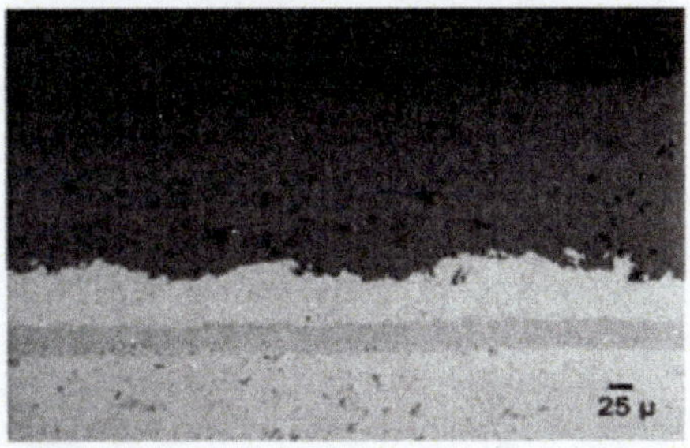

Fig. 4 Laminar structure of APS [18]

in a closed chamber that is free from air is known as vacuum spraying coating, and it is done at low pressures having range 0.1–0.5 atm [10] but for oxidation sensitive material like titanium, VPS is preferred over APS.

Merits. The coating obtained is dense and strongly adhesive in nature [3]. The porosity in case of APS is comparatively low that is only 10–20% [18]. Also, the coating is cheap enough that there is a less effect over the price of the component. The microstructure that is obtained by this coating is laminar in nature as shown in Fig. 4.

Demerits. The thermal conductivity of APS coating is comparatively low that is of the range of 0.8–1.1 W/m K [19]. Also, there is a problem of the oxidation of the metallic powder in case if the material of the coating is metallic in nature.

2.2 Electron Beam-Physical Vapor Deposition (EB-PVD)

The coating technique which is responsible for the deposition of any type of coating material is electron beam-physical vapor deposition (EB-PVD) [20]. In engines and aero-engines, EB-PVD has found a wide application, and also, this coating is extensively used in the gas turbines [18]. The arrangement for the setup used in the EB-PVD techniques is strenuous as compared to any other technique. An electron beam is originated by an electron gun at the high temperature of 2000 °C and the gun can be of type of electromagnetism deflection gun or pierce gun [20]. The beam

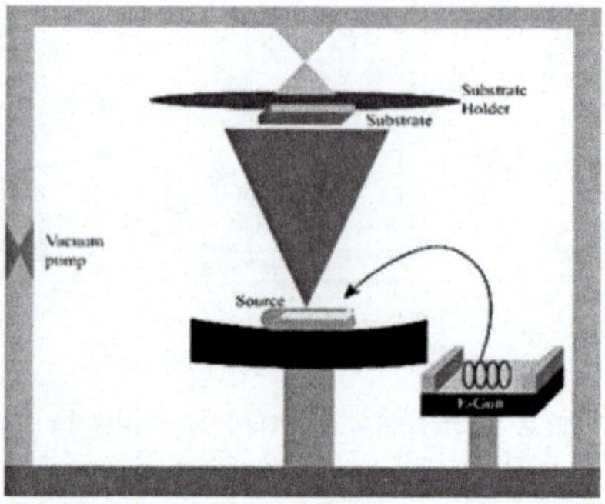

Fig. 5 Schematic diagram of EB-PVD

of highly accelerated electrons is directed toward material in the ingot, then there is melting of the material that is going to deposit on the metallic substrate in the form of vapors. The system of the EB-PVD techniques is having guns that is four in number from which one gun is used to raise the temperature of the metallic substrate and remaining three are there to heat the material carried in three ingots. To obtain the uniform thickness of the coating, two sample fixtures are utilized that also distributes the temperature uniformly over the substrate. Also, one pre-vacuum chamber and one simple chamber are there in the arrangement. The complete device and schematic diagram are shown in Fig. 5.

Merits. The surface roughness of the coating obtained by this method is of the order of 1–2 μm which says that the smooth surface of the coating is obtained [20]. The rate of deposition is high, and hence, the adhesion is strong as compared to other techniques due to the fact that process parameters of the system can be handled and controlled easily [20]. The life of this coating is more as compared to that of APS [18]. The microstructure that is obtained by this coating is stress concord at high rates which is columnar in nature as shown in Fig. 6.

Demerits. The inner surfaces of the specimen are hard to coat by this type of technique. If the pressure become less than 1.3×10^{-4} hPa, then formation of the cloud of vapor is there on the surface.

Fig. 6 Columnar structure of EB-PVD [18]

2.3 *High Velocity Oxygen Fuel (HVOF)*

As the name signifies, in HVOF coating, the use of fuel that can be hydrogen, propane, acetylene propylene, or kerosene along with the oxygen gas is used [2, 10] that is allowed to burn in the gun to obtain the flame under high pressure. Also, the kinetic energy of the material is very high due to the reason that the mixture of oxygen, fuel, and coating material is having the high velocity that has Mach number greater than 5 and hence considered as the supersonic flow [2, 10]. The material is in the semi-molten form that is further melted and passed through the nozzle [3]. The temperature of the flame is of the range of 2500–3200 °C [10].

The equipment of the HVOF consists of a gun having three inlets that is used to mix the fuel gas, coating material and oxygen gas in the mixing chamber and then the mixture is allowed to burn in the separate chamber termed as combustion chamber. The nozzle is used to direct the mixture after combustion toward the substrate. The schematic diagram is represented in Fig. 7.

Merits. High wear and corrosion resistance are obtained in HVOF technique, and also, the coating obtained by this method is highly dense that has porosity below 0.5% [2]. The microstructure that is obtained by this coating is lamellae elongated that is in the direction along the substrate as shown in Fig. 8.

Demerits. The coating is not economical but is expensive as compared to other techniques. Also, the narrow size particles have to be used in the process, and the completed process is automated.

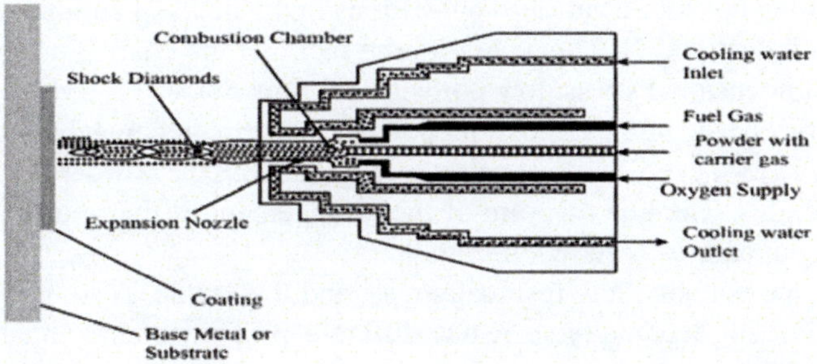

Fig. 7 Schematic diagram of HVOF

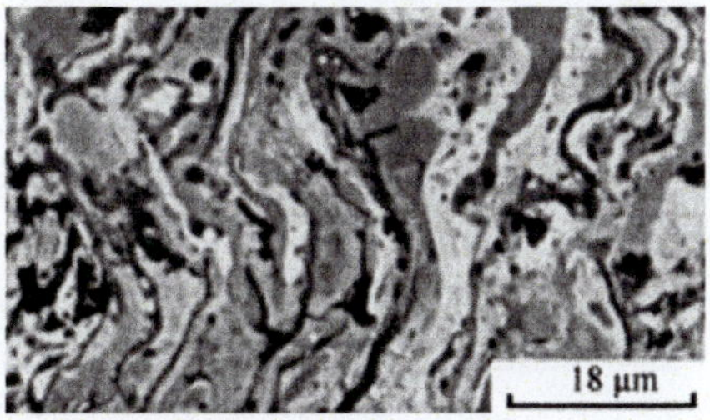

Fig. 8 Lamellae structure of HVOF [21]

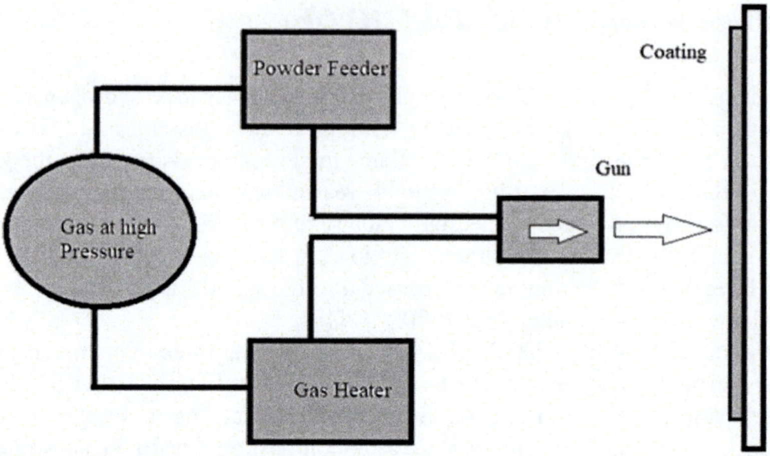

Fig. 9 Line diagram of apparatus of kinetic spray coating

2.4 *Kinetic Spray Coatings*

The unconventional technique used to deposit the coating material over the substrate in which the powdered coating material is accelerated with the help of the high velocity of a particular gas is termed as the kinetic spray coatings. The range of the velocity that is used to accelerate the material is of the order of 300–1200 m/s [22, 23]. The coating has found various applications in the fields of automobile, defense, and aviation industry [23]. This is due to the fact that the coating provided by this method has low residual stress, low porosity, and low oxide [22, 23].

The kinetic energy of the depositing powder is the main depending factor of this processes. Da Laval-type nozzle is being used to incident the powdered material over the substrate [23]. The line diagram of the arrangement of the setup of the kinetic spray coating apparatus is shown in Fig. 9.

Merits. Low porosity, low residual stress, and low oxide. Also, it is the type of solid spray coating technique so it has different properties as compared to other conventional coatings [24].

Demerits. Low ductility is there in the coating due to the fact of plastic deformation. Splat formation is also there in the coating due to the high kinetic energy [24].

2.5 *Comparison Between the Techniques of Coating*

The following table shows the comparison of different techniques of thermal spray based on the hardness, bond strength, and the thickness of the coting.

Table 1 Comparison of techniques of thermal spray coating process [3, 25]

Property	Coating type	APS	Electric arc	HVOF	KSC
Bond strength [MPa]	Ferrous	34>	41>	62	13–23
	Non	34>	41>	70	–
	Ferrous	–	–	62	–
	Ceramics	21>	–	–	–
	Carbides	55–69	–	83>	–
Hardness [HRC]	Ferrous	40	40	50	45–49
	Non	40	80 HRB	55	–
	Ferrous	–	–	60	–
	Ceramics	50–70	–	–	–
	Carbides	50–60	–	65	–
Coating thickness limitation [mm]	Ferrous	1.25–2.5	1.25–2.5	1.25–2.5	0.5
	Non-Ferrous	1.25–5	1.25–5	1.25–5	0.5
	Ceramics	–	–	1.25	0.5
	Carbides	0.4	–	–	–
		0.4	–	0.6	–

3 Conclusion

Various methods for the deposition of the thermal barrier coatings have been studied in this review that provides the basic idea about the various coating techniques (conventional and unconventional). It is found that HVOF techniques are the most suitable method to obtain the coating due to the following reasons:

- The coating obtained by HVOF is dense as compared to APS and EB-PVD as its porosity below 0.5%.
- From Table 1, it can be seen that the bond strength of HVOF is more than that of other coating techniques (APS, EB-PVD, and KSC) that is 62–83 MPa or more than that in case of carbides.
- The value of the hardness is more in case of HVOF that is of the order of 50–65 HRC as compared to other three coating techniques.
- The thickness for APS, EB-PVD, and HVOF coatings is comparatively equal that is 1.25–2.5 mm in case of ferrous material but the coating thickness in case of KSC is comparatively less that is of the order of 0.5 mm. So, it was observed the HVOF has comparatively more merits than other techniques of coating.

References

1. Sharma P, Dwivedi VK, Kumar D (2020) A review on thermal barrier coatings (TBC)—usage and effect on internal combustion engine. Advances in Fluid and Thermal Engineering. pp 77–85. LNME Proceeding ISBN: 978-981-16-0159-0.
2. Sankar V (2014) Thermal barrier coatings material selection, method of preparation and applications-review. Int J Mech Eng Robot Res 3(2). ISSN: 2278-0149

3. Barnwal S, Bissa BC (2015) Thermal barrier coating system and different process to apply them—a review. Int J Innov Res Sci Eng Technol 4(9). ISSN: 2319-8753
4. Garcia JRV, Goto T (2003) Thermal barrier coating produced by chemical vapor deposition. Sci Technol Adv Mater 4:397–402
5. Goto T (2005) Thermal barrier coating deposited by laser CVD. Surf Coat Technol 198:367–371
6. Beg RA, Bose PK, Ghosh BB, Banerjee TKr, Ghosh AKr (1997) Experimental investigation on some performance parameters of a diesel engine using ceramic coating on the top of the piston. ISSN: 0148–7191
7. Stover D, Funke C (1999) Directions of the development of thermal barrier coatings in energy applications. J Mater Process Technol 92–93:195–202
8. Padture NP, Schlichting KW, Bhatia T, Ozturk A, Cetegen B, Jordan EH, Gell M, Jiang S, Xio TD, Strutt PR, Garcia E, Miranzo P, Osendi M (2001) Towards durable thermal barrier coating with novel microstructure deposited by solution-precursor plasma spray. Acta Mater 49:2251–2257
9. Ghosh S (2015) Thermal barrier ceramic coatings—a review (Chap 5). http://dx.doi.org/https://doi.org/10.5772/61346.
10. Amin S (2016) A review on thermal spray coating processes. Int J Curr Trends Eng Res 2(4):556–563
11. Crawmer DE (2001) Plasma spray coatings. Encycl Mater: Sci Techol (2nd edn). 7035–7040
12. Das D, Majumdar G, Sen RS, Ghosh BB (2014) The effects of thermal barrier coatings on diesel engine performance and emission. J Inst Eng (India): Series C 95:63–68
13. Abbas SM, Elayaperumal A (2019) Experimental investigation on the effect of ceramic coating on engine performance and emission characteristics for clear production 214:506–513
14. Kumar D (2018) Study on the tensile and the fatigue behavior of air plasma sprayed YSZ TBC systems. J Metallic Mat Res 1(1):27–31
15. Vibha, Dwivedi VK, Kumar D (2019) Nano-Microstructural characterization of IN718 with thermal barrier coatings after thermal shock tests. Int J Innov Technol Exploring Eng (IJITEE) 8(5):36–41
16. Heimann B (1996) Plasma spray coating (Chap 3). VCH Publishers, New York
17. Bakan E, Vaben E (2017) Ceramic top coats of plasma-sprayed thermal barrier coatings: materials, processes, and properties. J Therm Spray Tech 26:992–1010
18. Beele W, Marijnissen G, van Lieshout A (1999) The evolution of thermal barrier coating-status and upcoming solutions for today's key issues. Surf Coat Technol 61–67
19. Xu J, Wu J (2011) New materials, technologies and processes in thermal barrier coating (Chap 15). Woodhead Publishing Limited
20. Zhang D (2011) Thermal barrier coatings prepared by electron beam physical vapor deposition (EB-PVD) (Chap 1). Woodhead Publishing Limited
21. Ji G-C, Li C-J, Wang Y-Y, Li W-Y (2006) Microstructural characterization and abrasive wear performance of HVOF sprayed Cr_3C_2–NiCr coating. Surf Coat Technol 200:6749–6757
22. Van Steenkiste TH (2001) Kinetic spray: a new coating process. Key Eng Mater 197:59–86. ISSN: 1662-9795
23. Lee C, Kim J (2015) Microstructure of kinetic spray coatings: a review. J Therm Spary Technol 24:592–610
24. Karthikeyan J (2014) The advantages and disadvantages of the cold spray coatings process. Fundam Appl 62–71 (Woodhead Publishing)
25. Moridi A, Gangaraj SMH, Vezzu S, Guagliano M (2014) Number of passes and thickness effect on mechanical characteristics of cold spray coating. Proc Eng 74:449–459

A Review on Post Additive Manufacturing Techniques to Improve Product Quality

Bhaskar Pandey and Rupesh Chalisgaonkar

Abstract Since past three decades, direct printing of parts for various applications by the use of 3D-printing technology has gathered a lot of attention. However, many issues remain with the widespread applications of 3D-printed materials such as regulatory issues, sterile environment for part production, and the improvement of material properties with the desired structure. This article focuses on review of various post additive manufacturing processes to improve mechanical properties such as storage modulus and Young's modulus and surface roughness. A heat treatment process after 3D printing was also discussed as potential method to achieve better mechanical properties by optimizing the crystalline structure of material. It was also reviewed that traditional AM technique produces products at a rapid rate and with minimum waste of materials and post processes could improve the surface finish as well as structure with higher strength in three dimensions.

Keywords Additive manufacturing processes · 3D printing · Mechanical property · Surface finish heat treatment

1 Introduction

Additive manufacturing (AM) technologies are nowadays have importance in industries (construction, prototyping and biomechanical) and in research field. Rapid prototyping which involve possibility of realizing objects of complex shapes is shifting conventional manufacturing techniques to produce customized products. 3D manufacturing system has benefits over conventional techniques in terms of overproducing products with complicated profile, high precision and accuracy, material savings, design flexibility and personal customization. Fused Deposition Modeling (FDM) is one of the most common AM methods due to user handy, economic and able to use conventional thermoplastic polymers such as PLA, ABS, and Nylon [1]. Three-dimensional (3D) printing technology has also potential to use products

B. Pandey (✉) · R. Chalisgaonkar
KIET Group of Institutions, Ghaziabad, India

A. K. Dubey et al. (eds.), *Recent Trends in Industrial and Production Engineering*, Lecture Notes in Mechanical Engineering,
https://doi.org/10.1007/978-981-16-3135-1_2

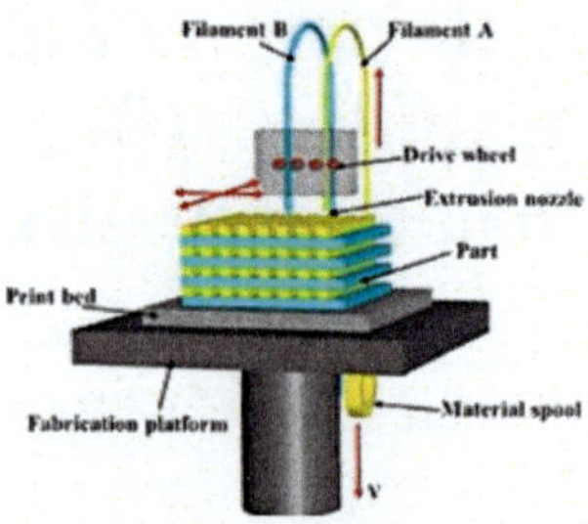

Fig. 1 Schematic diagram displaying fused deposition modeling process [2]

in biomedical field including engineering scaffolds, implants, printing tissues and organs, etc. The 3D printing has reduced the additional expenditure involved in the process of developing a product. FDM method is very common method of 3D printing which uses a continuous filament of a thermoplastic polymer to print layers of materials (Fig. 1a).

The filament is heated at the nozzle so that material will convert into semi-liquid state and finally extruded on the platform in a manner of layer by layer deposition to transform into final product. Material should be having higher degree of thermoplasticity so that it could fuse easily during printing. The main process parameters which affect the printed part's mechanical properties are thickness of the layer, width and filament orientation and air gap (in between or even the same layer). FDM process is economic, rapid and simple process but having disadvantages such as poor mechanical properties and poor surface quality. This paper reviews the various researches done to improve the mechanical properties and surface finish after 3D-printing process [2].

2 Temperature Effect on 3D-Printed Parts

Perfect 3D-printing temperature for material used such as PLA (polylactic acid) does not exist and various trial and errors are carried out to achieve the perfect PLA print temperature. Printing temperature impacts on the quality and effectiveness on 3D-printed product. An optimal range of temperature should be selected before starting 3D-printing process. Coppola et al. [1] investigated the prominent effects of temperature in the 3D printing of PLA/clay nanocomposites in additive manufacturing applications.

The two specimen names as Dog-bone and prismatic specimens were manufactured using the technique of fused deposition modeling at three varying temperatures. The temperature was increased in certain range from melting temperature to higher temperature as (185–200–215 °C) for PLA 4032D and (165–180–195 °C) for PLA 2003D as raw materials. The specimens were further tested for tensile strength according to the varying nozzle temperatures. The other mechanical properties such as young's modulus (E), elongation at break (εb) and tensile strength (σ) were also measured. It was found that the samples with nanocomposites have higher modulus of elasticity as compared to pure samples of polyacitic acid (PLA) for all the printing temperatures in his experiments. An embrittlement was found to be increased as nozzle temperature increases in given range for both PLA/nanocomposite specimens printed by FDM technique. Storage modulus variation was studied in defined range of printing temperature of PLA 4032D and PLA 2003D. The study revealed that storage modulus starts to decrease with rising print temperature for PLA 4032D while increases at increased temperature. The study revealed that printing temperature should be selected in correlation with the melting temperature [1].

It was also observed that the perfect adhesion of any printed sample to its printing bed was also affected by temperature variation and was achieved by heating the printing bed to temperatures slightly above the TG of the filament material for materials such as PLA. The simple reason behind this was that when the temperature is increased above the filament's TG it leads to a reduction of the surface tension between the printing bed and the printing material apart from it larger contact area is created which gives better adhesion between the bed and the filament. Strong increase in the adhesion force was also reported as printed bed temperature increases. It was observed that higher printing temperature generates a contact angle of low value as compared with the samples at lower temperature which results in more adhesion [3, 4].

3 Reason of Porosity and Affective Measures

In additive manufacturing density is defined as the pores development during the layering of the product. Although maximum number of products produced generally attain a density over 99.5% after 3D printing but repeatedly obtaining same density in manufactured products is undoubtedly challenging. The problems occurred due to porosity in a material was that it adversely affected the strength and reduced the resistance to impact of the product manufactured. Apart from it due to high porosity cracks, it generates or propagates faster whenever stress are generated during the life cycle of the product leading to its complete failure. Therefore porosity is a negative factor for finished product in 3D printing. Several researches have been done to minimize the porosity of the material.

Internal pores might be existing in 3D-printed specimen. X-ray computed tomography (XCT) technique might be useful to find out the details such as size, shape, and porosity. It was evident that the main parameters that were influenced the porosity of 3D-printed products were the thickness of the wall of specimen and extrusion thorough nozzle and G-code optimization of these parameters. It was found that larger the extrusion multiplier the porosity reduced significantly consistently. Whereas Wall thickness in the 3D-printed process allowed for an diagonal internal filling.

Further in some studies, the effect of porosity and crystallinity was studied in 3D-printed samples. It was found by specimen morphology that the pores intensity is more in the top and middle layer of 3D-printed specimen. It was suggested that thinner samples in conjunction with low build temperature lead to improved mechanical properties with higher degree of orientation [5].

Porosity is equally important in additive manufacturing or 3D printing of biomaterials as porosity along with size of pore of biomaterial scaffolds are crucial parameters which affect the quality and rate of bone formation. Lower porosity stimulates the process of osteogenesis by reducing cell proliferation phenomena and enhancing cell aggregation. Enhanced porosity tends to improve the growth in bone. Size of the pore shows its affect on the progression of osteogenesis due to vascularization, major experimental results show that small pores favored hypoxic conditions and induced osteochondral formation before osteogenesis and on the other hand large pores had been well-vascularized lead to direct osteogenesis without osteochondral formation [6].

Increased porosity in controlled manner of 3D-printed parts may be beneficial in case of biomedical implants. Minas et al. [7] incorporated large pores on top layer which resulted in better healing process. 3D-printed biocomposites consisting of higher porosity introduces a hygroscopic property which increases to retain the water.

4 Surface Roughness Affects and Measures

Surface roughness is defined as the variance or irregularities in a surface topology. Surface roughness details and specifications are important and crucially required information in any engineering design as it determines the life of the product and usability. Metal Additive Manufacturing (AM) processes although takes some care of the surface roughness but usually fails to meet the precise surface roughness requirements. This makes it necessary for some expensive post-processing techniques such as machining or polishing. Surface roughness capabilities of metal AM, post-processing techniques and their associated time and cost should be considered to select the optimal manufacturing workflow.

Surface roughness of Metal AM parts have three major contributors that are

- Surface irregularities due to layering effects and low process resolution,
- Granular micro-surface textures from melting and binding powder feedstocks,
- Support structures and the remnants and surface marks left by their removal.

The main process parameters that generally affect the surface roughness of AM product are speed of printing, power, built part, and cooling rates.

Several studies show that the layer height with wall thickness is critical factors to be selected for optimum surface roughness. The surface roughness is decreased when height of the layer and product wall thickness are increased simultaneously or individually in FDM technique. It was recommended to use lower levels of layer height, staircase effect and also wall thickness to reduce surface roughness. The parameter selection should be also based on the size of the nozzle extruder [8].

Improving the surface roughness can be really easy and can be done without making any major changes in the main processing or post-processing techniques simply by optimizing the design which includes geometry and orientation of the parts and supports. Therefore implementation and making of most optimized design is the most effective way to control surface roughness.

A new post-processing method has been reported for 3D-printed thermoplastic parts which are also called constrained remelting. The improvement in the surface roughness and mechanical strength of the 3D-printed parts was reported. A 3D-printed PLA sample was inserted in a profiled mold insert with a negative shape and was then heated to near the melting temperature with additional thickness. Remelting conditions, the remelting temperature and initial thickness were investigated in terms of the surface roughness and tensile strength. The results showed that remelting at 160 °C along a 4.0 mm thickness could improve surface finishing. [9, 10].

5 Warping

The major effect of wrapping is the upliftment or rising of the deposited metal at the ends this is caused due to rapid cooling of the material. During rapid cooling the material contracts quickly and results in the above problem. Warping also results in formation of cracks in the print. The main reason for this problem is that the temperature difference between the layers of extruded product is significantly large, which creates tension within the product and consequently the lower layer start to lift up or drag. Easy solution to this problem is to just keep the entire printed product at the same temperature. But this is a nearly impossible task therefore several ways have been devised to overcome wrapping such as adjusting bed temperature, using bed adhesion tools, and setting optimum enclosure temperature.

Alsoufi and Elsayed [11] found that extruded PLA + filament deposited onto the non-heated platform by moving a printing head in a user-defined pattern, such as to achieve the desired final 3D shape. It was evident that the optimum value for temperature of the nozzle was nearly 220 °C with minimum warping deformation of 4.77 ± 0.12 mm at 15 mm/s printing speed. The accuracy of arrangement resulted in a small percentages error of about 4.55%. The result of the validation process showed that there was a significant improvement in warping deformation in each case by reaching 0.4% error.

It was also found that the printed products with 90° edge also showed less wrapping deformation. Another study says that it is found effective to use a enclosure or enclosed 3D printers. Although it makes it a little costly but helps prevent defects like wrapping and clogging of nozzle therefore properly maintaining the enclosure temperature should be considered as a crucial step.

Heating bed during printing is also one of the main solution to eliminate wrapping completely. The core idea is to keep the first layer heated and not let them cool quickly. This helps the initial layers to remain in contact with the build plate thus avoiding warping [12].

6 Quality Methods

Since roughness and layer line are mostly present of the FDM-based products therefore for smooth surface post-processing becomes crucial. Some of the post-processing techniques not only adds strength to prints but also helps to reduces the anisotropic behavior of it. One of the common methods is to annealing the produced part. Slavković et al. [13] studied the mechanical properties of 3D-printed polylactide acid SMP material. Annealing treatment and printing directions were considered and found to have significant impact on elastic modulus and ultimate strength of material. Annealing at 75 °C increased tensile and compressive strength compared to as-printed SMP tested at room temperature. It is probably due to annealing effects resulting in residual stress relief and crystallinity stronger layer bonding which causes changes in the properties of 3D-printed products.

Electrode position-based AM technique improves the deposited metal qualities and eliminates limitations and improving horizontal and vertical resolution. Electrode position has the quality of producing films in vertical resolution at a sub-nanometer range [14]. Electroplating is also considered as a good post-processing technique as it can improve the strength of the product but also gives a fined surface finish enhancing the look of the final product. Materials that can be easily electroplated include ABS.

It was found that if titanium alloy is produced using 3D printing in traverse direction as compared to longitudinal direction then mechanical properties such as tensile strength and ductility could be improved [15–17]. Same observations were identified for alloys, ceramics and polymers [17–19].

7 Crack Detection in 3D-Printing Process

Cracking is caused due to deformation in higher layers after printing process. The stress caused between two layers tends to deform the layers due to which crack is initiated in the object.

Patterson AE [20] studied the most affected areas on which cracks could initiate and hence grow in 3D-printed materials.

(1) Cracking caused by faulty print process.
(2) Cracks initiate in the shell and proceeds toward relatively low dense areas of 3D-printed part. The other attributes caused of cracking is also due to rapid material cooling and residual stresses induced.
(3) Cracking and delamination of parts have been also reported which are printed into full density parts. This is generally caused by residual stresses and concentrations in the part after printing. Effect of residual stresses on full density parts is prominent more severely than partially filled parts.

Materials having higher rates of shrink develop more cracks after cooling. It was also found that the cracks stay without further escalation till any external source

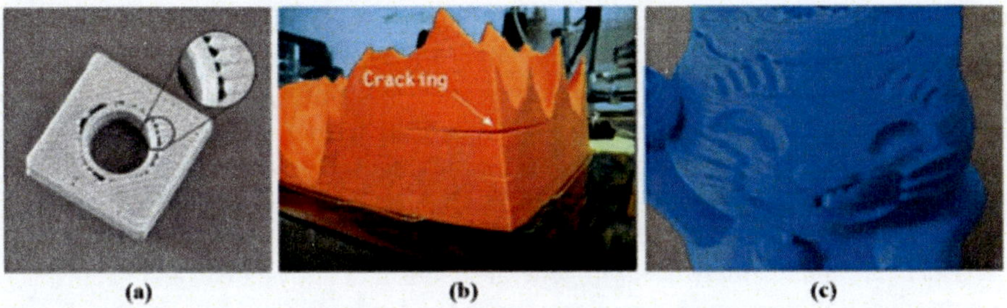

Fig. 2 **a** Surface defect crack. **b** Shell defect crack. **c** Delamination [20]

acts or imparts energy on the product. Zolfagharian [21] did experimentation using centrally positioned notches with Nylon 12 filament and PA12 nylon powder to produce 3D-printed specimens. This printing technique was found to be effective for improving tensile behavior. MJF (multi-jet fusion) and FDM (Fused Deposition Modeling) was compared in terms of structural integrity, elongation and toughness. FDM process was found to be most suitable as compared to MJF considering all mechanical properties.

Layer height should be 20% smaller than the nozzle diameter selected as reported in one study. If the layer height exceeds the limit then improper bonding of each layer of plastic may result. For avoiding cracks during 3D-printing process print temperature should not be too low and print speed should be controlled properly as it should not be in higher range [22] (Fig. 2).

8 Material Selection

The applications of additive manufacturing are widespread and escalating at a rapid rate; therefore, the choice of material for different application becomes important for a efficient three-dimensional printing process. Generally, 3D-printing process begins with melting of the metallic feedstock for which usually an electron beam or a laser beam is used and after that a solid part is obtained layer by layer. Many materials such as alloys of steel, aluminum, magnesium, cobalt, nickel and titanium along with pure titanium are considered good for a 3D-printed product. Titanium and its alloys are used widely in many industries and workplaces such as aerospace and biomedical fields because of its high performance but the machining cost of titanium is high. Various steels such as austenitic stainless steels, precipitation harden able stainless steels, maraging steels are commonly used in AM. These alloys are used to obtain high strength and wear or scratch resistance that is hardness conditions. Aluminum compared to titanium is used very less in 3D manufacturing but they are easily machinable and cheaper compared to titanium. Due to high thermal conductivity of aluminum, the residual or internal thermal stresses are reduced that allow faster production. Further, magnetic properties are required in areas like aerospace, etc.;

therefore, such material are also prepared in 3D printing, and studies are being done on magnetic alloys such as NiFeV and NiFeMo. Although a vast variety of metals and their alloys are being used in this printing, majority of the alloys do not fit good in the process of 3D printing due to melting and rapid solidifications. Additive manufacturing is now also used in the construction process but here it depends on a large nozzle size and concrete properties like good extrusion strength and cohesion among concrete particles. Although a vast variety of material are used in AM process, polymers are most widely used. In general Metals and alloys are used in aerospace, automotive. Polymers and composites are used in Aerospace and Automotive, Sports, Medical, Architecture, Toys and biomedical. Ceramics in chemical industries and aerospace and concrete in infrastructure and construction [2, 23, 24].

9 Conclusion

3D-printing technique has various advantages over the convention methods like faster production, less tool requirement, accuracy and it is much economic to use. Although FDM generally used ceramics as printing ink but nowadays metals, alloys and even concrete are used for the same. The process suffers from main disadvantage of poor mechanical properties and surface finish. It was found that printing temperature was playing a important part to attain higher mechanical properties such as Young's modulus (E), net deformation at break (εb), and elastic or tensile strength (σ). Even selecting a temperature higher than the TG of the material resulted in a better material and bed adhesion which is desiray. If larger extrusion multiplier values (sometimes as large as possible) are used then porosity could be reduced. In 3D printing, internal diagonal filling should be allowed in wall thickness. Some research recommends a low or reduced value of wall thickness, a decreased height and stair casing effect reduce surface roughness. Remelting is another post-processing method which has been found to improve the surface roughness and mechanical strength of the 3D-printed parts. Several methods such as adjusting bed temperature, using bed adhesion tools, and setting optimum enclosure temperature have been found to overcome wrapping problem. Annealing treatment and printing directions were found to be vital important for improving mechanical properties such as elastic modulus and ultimate strength of material. The various research conducted focuses on to improve the surface and mechanical properties of 3D-printed product. Future research could lead to a stage where we can perform AM in a manner without producing the common defects and even eliminating processes like heat treatment which will definitely reduce the cost of production and eventually reduce the production time. But for now, working with metals, ceramics or alloys is still a challenge with the present technology of AM.

References

1. Coppola B, Cappetti N, Maio LD, Scarfato P, Incarnato L (2018) 3D Printing of PLA/clay nanocomposites influence of printing temperature on printed samples properties. Materials 11:2–17. https://doi.org/10.3390/ma11101947
2. Ngo TD, Kashani A, Imbalzano G, Nguyen KTQ, Hui D (2018) Additive manufacturing (3D printing): a review of materials, methods, applications and challenges. Composites B 143:172–196
3. Spoerk M, Gonzalez-Gutierrez J, Sapkota J, Schuschnigg S, Holzer C (2017) Effect of the printing bed temperature on the adhesion of parts produced by fused filament fabrication. Plast Rubber Compos 47(1):17–24. https://doi.org/10.1080/14658011.2017.1399531
4. Sun H, Wang SQ (2012) Shear and extensional rheology of entangled polymer melts: similarities and differences. Sci China Chem 55:779–786
5. Liao Y, Liu C, Coppola B, Barra G, Maio LD, Incarnato L, Lafdi K (2019) Effect of porosity and crystallinity on 3D printed PLA properties. Polymers (Basel) 11(9):1487. https://doi.org/10.3390/polym12010142
6. Karageorgiou V, Kaplan D (2005) Porosity of 3D biomaterial scaffolds and osteogenesis. Biomaterials 27:5474–5491. https://doi.org/10.1016/j.biomaterials.2005.02.002
7. Minas C, Carnelli D, Tervoort E, Studart AR (2016) 3D printing of emulsions and foams into hierarchical porous ceramics. Adv Mater 28(45):9993–9999
8. Pérez M, Medina-Sánchez G, García-Collado A, Gupta M, Carou D (2018) Surface quality enhancement of fused deposition modeling (FDM) printed samples based on the selection of critical printing parameters. Materials (Basel) 11(8):1382. https://doi.org/10.3390/ma11081382
9. Cho KL, Liaw II, Wu AHF, Lamb RN (2010) Influence of roughness on a transparent superhydrophobiccoating. J Phys Chem C 114:11228–11233
10. Kim HC, Kim DY, Lee1 JE, Park K (2017) Improvement of mechanical properties and surface finish of 3d-printed polylactic acid parts by constrained remelting. 8(12):1199–1203. https://doi.org/10.5185/amlett.2017.1686
11. Alsoufi MS, Elsayed AE (2017) Warping deformation of desktop 3d printed parts manufactured by open source fused deposition modeling (FDM) system. Int J Mech Mechatron Eng 17(4):7–16
12. https://manufactur3dmag.com/common-problems-in-3d-printing-how-to-resolve-them-part-i/
13. Slavković V, Grujović N, Dišić A, Radovanović A (2017) Influence of annealing and printing directions on mechanical properties of pla shape memory polymer produced by fused deposition modeling. In: 6th International Congress of Serbian Society of Mechanics Mountain Tara, Serbia, pp 1–8
14. Braun TM, Schwartz DT (2016) The emerging role of electrodeposition in additive manufacturing. Electrochem Soc Interface 25(1):69–73
15. Carroll BE, Palmer TA, Beese AM (2015) Anisotropic tensile behavior of Ti–6Al–4V components fabricated with directed energy deposition additive manufacturing. Acta Mater 87:309–320
16. Mühler T, Gomes CM, Heinrich J, Günster J (2015) Slurry-based additive manufacturing of ceramics. Int J Appl Ceram Technol 12(1):18–25
17. Niendorf T, Leuders S, Riemer A, Richard HA, Tröster T, Schwarze D (2013) Highly anisotropic steel processed by selective laser melting. Metall Mater Trans B 44(4):794–796
18. Cooke W, Tomlinson RA, Burguete R, Johns D, Vanard G (2011) Anisotropy, homogeneity and ageing in an SLS polymer. Rapid Prototyping J 17(4):269–279
19. Guessasma S, Belhabib S, Nouri H, Hassana OB (2016) Anisotropic damage inferred to 3D printed polymer using fused deposition modelling and subject to severe compression. Eur Polymer J 85:324–340
20. Patterson AE (2016) Crack propagation in 3-D printed PLA: finite element modeling, test bed design, and preliminary experimental results. A Technical Report, University of Illinois at Urbana-Champaign

21. Zolfagharian A, Khosravani MR, Kaynak A (2020) Fracture resistance analysis of 3D-printed polymers. Polymers 12:302. https://doi.org/10.3390/polym12020302
22. https://www.gearbest.com/blog/how-to/how-3d-printing-fills-the-gap-between-the-outline-2992
23. Wach RA, Wolszczak P, Adamus-Wlodarczyk A (2018) Enhancement of mechanical properties of FDM-PLA Parts via thermal annealing. Macromol Mater Eng 303:1800169
24. Song Y, Li Y, Song W, Yee K, Lee KY, Tagarielli VL (2017) Measurements of the mechanical response ofunidirectional 3D-printed PLA. Mater Des 123:154–164

A Review on Various Types of Burr Formation in Drilling

Vijay Kumar Dwivedi and Anas Islam

Abstract The process of drilling is commonly employed in the processes involving the complex manufacturing processes. Burr exists in almost all these process whether it be drilling, milling or any kind other kind of manufacturing operations. The burrs that are generated on any of the component (part) may lead to several unwanted effects: practically like not having any proper contact between the members that are the carriers of currents and also the irregular arrangements between the contacted surfaces of the components. Burrs could also be injurious while performing some machining operations since they can be responsible for creating the wear of groove of the cutting edges. Subsequently, it is necessary to understand those elements that influences the arrangement of burrs at the exit of the openings in drilling operations in order to decrease the burr size at the site of their generation. The central target of this review research work is to complete the previous several years in this significant theme. The paper remember the conversation for the burr development in different drilling procedure improvement, impacted by the cutting conditions and the drill geometry during penetrating of different sorts of work materials, utilizing turn drills. The drilling improvement is performed to decide the ideal benefits of cutting pace, feed, point edge and lip angle of clearance for a predefined drill distance across that at the same time limit burr size, to be specific, burr height (tallness) and thickness of the burrs.

Keywords Burr formation · Process parameters · Uniform · Transient · Crown

1 Introduction

The formation of burrs in the processes like milling, drilling, grinding and turning or engraving is one of the main issues that is encountered while going through (manufacturing) the precision engineering components and producing in the in bulk quantity. Burrs are of several different kinds that minimize the quality and also affect

V. K. Dwivedi · A. Islam (✉)
IET Department of Mechanical Engineering, GLA University, Mathura 281406, India

A. K. Dubey et al. (eds.), *Recent Trends in Industrial and Production Engineering*, Lecture Notes in Mechanical Engineering,
https://doi.org/10.1007/978-981-16-3135-1_3

the precision of the products. Thus, it will further increase the deburring costs up to a significant amount that is incurred in the overall costs of the manufacturing [1]. Exit burrs formation on the edges of parts while drilling, contains several unwanted features with respect to the quality of the product and also affect its functionality. No tools for deburring are available when the formation of the exit burr takes place inside a cavity. Some special types for tools should be used in order to do deburring in such places; this will increases the cost involved in deburring. Generally, the process of deburring is done by manual methods as there are several problems arise for making an automation system of removing of burrs (deburring). The difficulties that are involved in the automation of deburring are: the automation may cause the damaging of the edges of the work piece; thus, the work piece might be rejected and the overall incurred cost may get increased. It might likewise cause significant expense in edge completing of accuracy parts. The burrs are undesirable as they minimize the finishing and also interfere in assembling the subsequent parts; such kinds of interference may also results in the misalignment and can cause the jamming when put into practice [2]. There are certain reliability issues that degrade the precision of these involved parts. In present day manufacturing processes, accuracy in the precision of parts requires more attention regarding both of surfaces and dimensions of diameter the across, with exceptionally close resiliencies. High calibre and decisively items ought to be made as per the designed plan measurement with low assembling and creation costs. To accomplish these objectives, the assembling procedure ought to be comprehended and its boundaries should be limited [3, 4]. The traditional ways of manufacturing like metal forming, machining, castings and injection mouldings. Drilling is considered as one of the widely practiced operation among several methods involved in machining. Industries like automotives and aerospace commonly uses the process of drilling. Burr is responsible for creating the dimensional errors that take place mostly while the drilling operation is being performed. Nearly all the machining processes give rise to the formation of "burrs". The burr may be defined as the material that is extended out of the work piece while doing the machining operations. Burr may also be called as the material that is deformed plastically; it is set-up on the edges of the parts of the work piece while doing the shearing or cutting operations. As soon as the operation of drilling is done, it can be concluded that these kinds of burrs formed immediately on the cutting edge onset as soon as the edge of the chisel and the work piece surface comes in contact. The propagation and formation of these burrs is in the circumferential direction as the bit is travels inside the work piece. The analysis of the area measurement of burr based on its surface is shown as in Fig. 1.

The classification of burr based on the drilling speed, size of bit and cooling medium is as summarized in the Table 1. The images of crown burrs, rolled over burr and uniform burrs are as shown by Fig. 2a, b, and c, respectively.

The process parameter that affects the formation of burrs are summarized in Table 2.

A summary of these reviewed papers are briefed as in Table 3.

Table 4 shows the classification of burr as done by Kim et al. First type denotes the uniform burr which is having smaller height of the burr while the type II corresponds

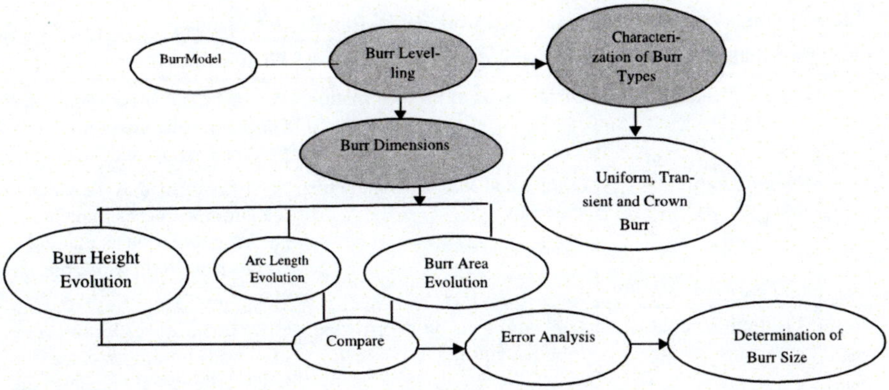

Fig. 1 Flow chart for the analysis of area of burr measurement

Table 1 Type of burr produces with a bit size of 3 mm as per the different process parameters [5]

S.No.	Speed (in rpm)	Coolant type	Type of the burr	Remarks
1	1400–1500	Air	Uniform	Taken by digital microscope
2	1500–1600	Water	Uniform	Taken by digital microscope
3	1600–1700	Water	Crown type	Taken through photo profilometer
4	1700–1800	Air	Rolled over	Taken by photo profilometer

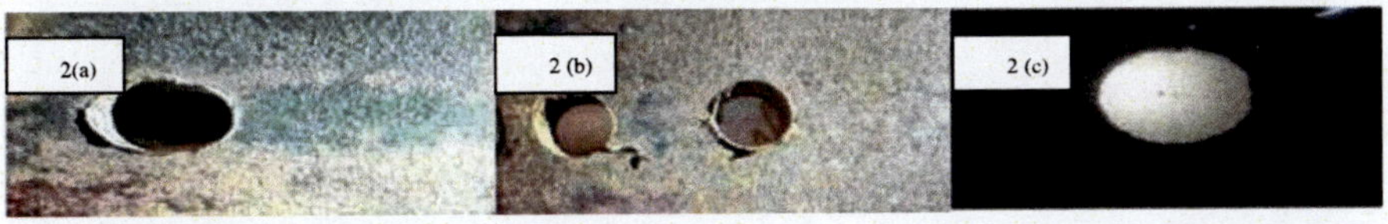

Fig. 2 Crown burr, rolled over burr, uniform burr taken through photo profilometer at 1678, 1750 and 1450 rpm, respectively

Table 2 Parameters affecting drilling burr formation [6]

S.No.	Category	Parameters
1	Process conditions [7–9]	Cutting speed, feed, use of coolant
2	Drill geometry [10, 11]	Point geometry, point angle, lip clearance angle, helix angle
3	Material properties [12]	Ductility, hardness, tensile toughness, strain-hardening characteristics
4	Others [13, 14]	Temperature dependence properties

Table 3 Summary of the previous research work on the Burrs

S.No.	Ref. authors	Methodology	Findings
1	Gillespie and Blotter	On the basis of their formation	Poison burr is predominate for entry and rolled over are predominant for exit burrs
2	Ko and Dornfeld	Orthogonal M/C tests inductile materials	A burr is formed as the value of the strain increases in negative shearing direction as the tool comes near to the work piece end
3	Hashimuraet et al.	Analysis of 3D burr formation "oblique cutting"	Exit burrs are smaller in case of oblique cutting as compared to the orthogonal one, but "side burrs" in orthogonal cutting are larger than orthogonal
4	Kim et al.	Classified burrs as per their applications	Three types of burrs were characterized like small uniform as type I, large uniform as type II and type III for the crown burrs
5	Dornfeld et al.	Burr formation while drilling for "Ti6Al4V"	The generated heat while the drilling operation is being performed is the major deciding factor for the final type of the burr formed
6	Kitajima et al.	Classified the exit burrs for drilling in Al-alloys	Highest value the temperature could be observed at the hole's centre; also its value decreases towards outside edges of the holes
7	Sofranest et al.	Shape of the drilling burrs are governed by various causes	Material of the work piece, geometry of tool (drill) and process working conditions are the most dominating factors
8	Gillispie	Ti alloys were used for characterizing the burr formations	Experiments were done by hand drill in uncontrolled and unknown feeding rate; feed rate is also one of the process parameters along with other process parameters
9	Charn and Dornfield	Edge breakout and formations of burrs	In case of the oblique cutting, no mathematical equations could be employed for the prediction of the burr formations

(continued)

Table 3 (continued)

S.No.	Ref. authors	Methodology	Findings
10	Sofronas and Taraman	Carried out theoretical models and also experimentations	Thickness of the exit burr might be reduced by increasing the angle of helix, clearance of the lip angle, decrement in the feeding rate
11	Stein and Dronfield	Found the sensitivity for different process parameters in drilling in case of the "Stainless steel 304L"	Burr height is influenced by increasing the feeding rate, speed of cutting and the wiring in the drill bit are likely to increase the height of the burr produced
12	Sugawara and Inagki	Gave insights about the quantity about burr formation	Burr quantity increases as the diameter of the bit decreases in the range "0.2–2.5 mm"; if the diameter of the burr below 0.2, then burr quantity would be supposed to decrease with decrease in the bit diameter
13	Min et al.	Gave results on burr height for "poly crystallined copper"	Suggested the burr height as a function of the orientation of the drill in case of the "poly-crystalline Cu". For one grain, the cutting mode may like as in ductile while for others, it might be like that of brittle material
14	Takzawa	Size of the burr produced	The burrs that are formed through drilling as produced by a drill having a nick on the cutting lips were found to be smaller one as compared to that of the burrs that are produced through the conventional burrs
15	Kim	Effect on burr on Ti alloys by different drill	Two different carbide drills were used in his experimentation work. No coolant was used in these experimentations. Only smaller burrs of type I and II are mostly formed

Table 4 Classification of burrs [6]

		Type I	Type II	Type III
Burr type		Uniform burr having a drilling cap		Transient, crown burr
Height of the burr in mm	AISI304L AISI4118	~0.15 ~0.18	0.15~1.1 0.18~1.0	(1.1~1.5) × (d/2)

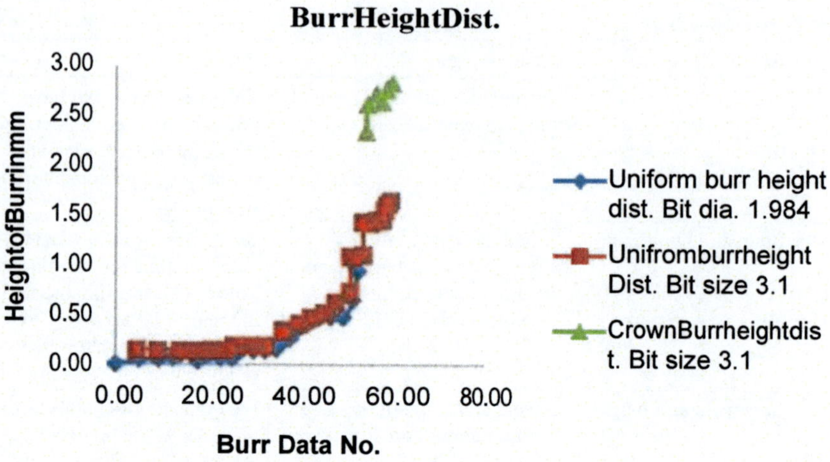

Fig. 3 Height distribution of burr according to burr data number

to higher burr height. These two types of the above-mentioned types of burrs do not depend upon the diameter of the drill. The type III is a crown burr which does not have any type of drilling cap, while its height depends upon the diameter of the hole.

The classifications of the burrs could vary as per the type of the need; e.g. when the aim is to resist the crown burr formation, type I and II could be mapped together. It is evident from the graphical representation as in Fig. 3 that the burr height distribution for the two drill bit diameters, viz. 1.98 and 3.1 mm, the height of the burr varies as suggested by Kim et al. in 2001.The height distribution of the burr formation has been briefed by the graphical format as in Fig. 3.

2 Results and Discussions

After a comprehensive survey in this survey, many researchers have provided as per their findings regarding the major parameters affecting the formation of the burrs, and in accordance with it, a summary of these parameters have been shown as a pie chart on the area characterizing the amount due to which the formation of the burrs gets affected. In the pie chart as discussed in Fig. 4 corresponds the parameters that are responsible for affecting the burr shape, size and height.

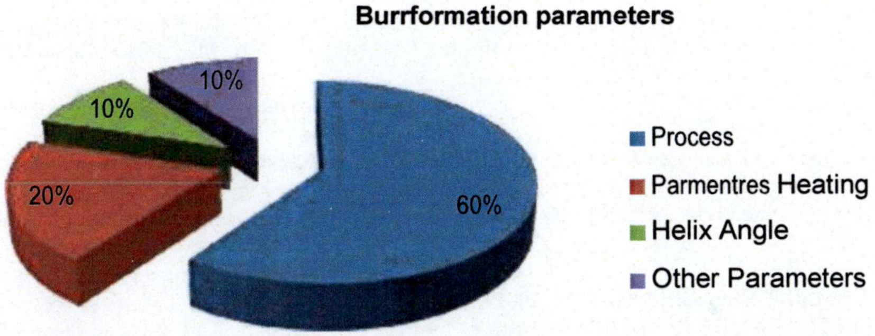

Fig. 4 Pie chart conceptualizing the parameters that influences the burr formation

3 Conclusions

- As found by an exhaustive review of the literature by a number of authors, the optimal value settings for the feed, clearance angle and cutting speed need to be find out that influences the burr height and its thickness. Thus, in order to minimize these unwanted burrs, the proper settings of these process parameters play a vital role.
- Some previous studies also give an insight as the minimization of burrs linked as an objective function that needs to be optimized (minimized). All the studies have shown the objective function in case for the minimization of burr is either to minimize its height or thickness but no study reveals the minimization of these two simultaneously.
- Optimization processes that are ason GA and Taguchis are quite robust in nature and also powerful ways of optimization that economically satisfies the requirement of problem (optimization) for problem solving and also are considered as a good optimization of designs.

References

1. Bahçe E, Özdemir B (2020) Burr measurement method based on burr surface area. Int J Precis Eng Manuf Green Technol https://doi.org/10.1007/s40684-020-00228-0
2. Gillespie LK (1999) Deburring and edge finishing handbook. Society of Manufacturing Engineers, Dearborn, Michigan
3. Min S, Lee D, Grave A (2005) Surface and edge quality variation in precision machining of single crystal and polycrystalline materials. LMA Annu Res Rep 2003–2004:52–62
4. Kim J, Dornfeld DA (2000) Development of a drilling burr control chart for stainless steel. Trans NAMRI/SME 28:317–322
5. Islam A, Dwivedi VK (2020) Effect of drilling speed, bit size and cooling medium on the burr structure for mild steel. Mater Today: Proc. https://doi.org/10.1016/j.matpr.2020.04.812

6. Kim J, Min S, Dornfeld D (2001) Optimization and control of drilling burr formation of AISI 304L and AISI 4118 based on drilling burr control charts. Int J Mach Tools Manuf 41(7):923–936
7. Ko SL, Dornfeld DA (1991) A Study on burr formation mechanism. Trans ASMS, J Eng Mater Technol 113:75–87
8. Kitajima K, Yamamoto A, Miyake T, Takazawa K (2005) Influence of work piece temperature on burr formation in drilling. J Jpn Soc Precis Eng 71(2):252–256
9. Min S, Dornfeld DA, Nakao Y (2003) Influence of exit surface angle on drilling burr formation. Trans ASME, J Manuf Sci Eng 125(4):637–644
10. Riech-Weiser C, Dornfeld DA (2005) Drilling burr control chart—adding a material property axis. LMA Annu Rep 2004–2005:19–21
11. Lin TR, Shyu RF (2000) Improvement of tool life and exit burr using variable feeds when drilling stainless steel with coated drills. Int J Adv Manuf Technol 16:308–313
12. Huang MF, Lin TR (2004) Application of grey-Taguchi method to optimise drilling of aluminium alloy 6061 with multiple performance characteristics. Mater Sci Technol 20(4):528–532
13. Ko SL, Chang JE, Kalpakjian S (2003) Development of drill geometry for burr minimization in drilling. Ann CIRP 52(1):45–48
14. Jean MD, Wang JT (2006) Using a principal components analysis for developing a robust design of electron beam welding. Int J Adv Manuf Technol 28:882–889

Application of 5s Methodology in a Small-Scale Enterprise: Case Study

Vijay Kumar Dwivedi, Anas Islam, and Aman Sharma

Abstract Small-scale industries or enterprises play a vital role in building up the economy and further enhancing the growth of any country. Likewise, there are several small-scale industries and startups that are currently active in our country. The countries like India where population holds the second place in the world small-scale enterprises are of the great importance. As far as the medium and the big industrial setups are concerned, there are always some methodologies that are adopted in order to enhance its growth and minimize the wastages. While in case of the small enterprises, there is no any as such provision for applying any lean manufacturing or management technique for enhancing the work culture and minimization of the wastages. In this regard, a very known 5s technique (lean manufacturing method) is applied in one of the small industries located in a district of Uttar Pradesh Aligarh. The name of the industry is "Reliable Tools," and it is known for manufacturing the locks and its parts. After applying this technique, the results and the final effects of this technique were analyzed based on the wastages of materials, time and other miscellaneous items. Finally, the results are concluded by comparing costs incurred before and after applying this 5s technique.

Keywords 5s methodology · Lean manufacturing · Small-scale industries

1 Introduction

The town of Aligarh has for some time been known all through the world for its lock industry. Truth be told/when the word Aligarh is stated, the unavoidable answer comes—"Gracious, the town of the lock. Also, why not? At the point when we took a gander at the present territory of Aligarh, we see that there are around 3000 units providing locks and their segments using upwards of one lakhs resident with a turnover of thirty crore for each year". The manufacturing of locks and their industrial setups is spread all over the city in small and big firms. As far as present work culture

V. K. Dwivedi (✉) · A. Islam · A. Sharma
IET Department of Mechanical Engineering, GLA University, Mathura 281406, India
e-mail: vijay.dwivedi@gla.ac.in

A. K. Dubey et al. (eds.), *Recent Trends in Industrial and Production Engineering*, Lecture Notes in Mechanical Engineering,
https://doi.org/10.1007/978-981-16-3135-1_4

is concerned, only big organizations take part in adopting the lean manufacturing techniques and attain good profits.

A good management will lead the industry/startups to a new height and also minimize excessive wastages.

5S is commonly the initial phase in wiping out squander (unnecessary items). 5S is the foundation of kaizen, a technique for nonstop improvement concentrated on certain core values, which include expanded productivity, enhance safety, less wastages and lesser lead time, etc. The 5S methodology is very well suited for a place for everything and everything in its place (PEEP) [2, 10, 11]. 5S is a waste reduction and efficiency enhancing framework that maintains a methodical work environment and uses visual pieces of information like signboards and charts to achieve increasingly reliable operational outcomes [1]. The word 5S consolidates five establishments and these can be expressed as Seiri (Sort) Seiton (Set), Seiso, (Shine) Seiketsu (Sustain) and Shitsuke (Standardize).

In 5s technique, less wastages and in turn high profitability are achieved [3–5]

- Execute the implementation with the guide of posters, banners, shadow boards, gadget holders and so on.
- Without the sustain column, the accomplishments of different columns would not keep going long [6].

Themethodology introduced by Hirano states that implementation should be articulated in such a way that the less complex and important frameworks should be enforced first. Hirano explains the series by a famous flow chart known as Hirano's 5s implementation as shown in Fig. 1.

2 5s Implementation

2.1 Seiri (Sort)

Seiri expects/tends to remove everything from the workplace that is not required for current process. The implementation of Seiri is executed in the following manner:

- The workplace must be divided into various zones depending upon the work they are performing with the main goal of including each and every piece of workplace in the endeavor at the phase such that all these zones are connected together. Allocate the coordinators for every one of these zones.
- Identify a red name (red tag) holding items. Red marked things are those that are viewed as pointless in a specific working area/department. The red label keeping region is the zone ensured for use in the substitution of red names that need further assessment. Define the red tag. A red tag is a paper name with relating information:
  - What's item?
  - How much quantity?

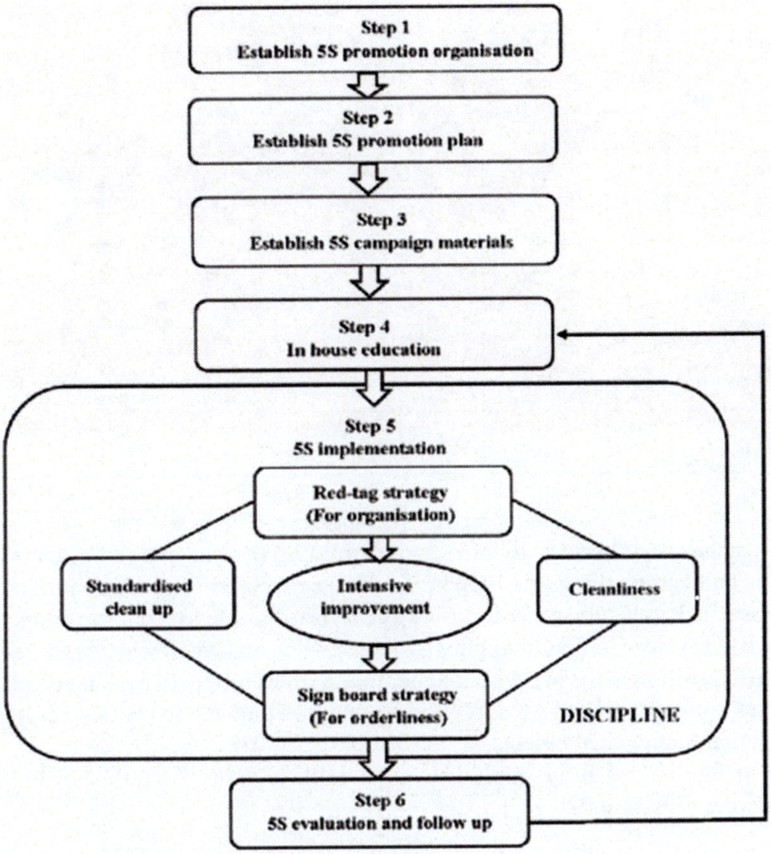

Fig. 1 Practical approach to incorporate 5s in an Industry [4]

- Why are they expelled (hurt, over a boundings, etc.)?
- What is the estimated cost?
- The area from which the thing is removed?
- Establish the repeat of red label checking; an example of red tag is shown in Fig. 2.

2.2 *Seiton (Set)*

It is indeed important to place the essential products in good alignment such that these could be quickly grabbed to be used. It is a performance analysis. It is indeed a matter as to how fast you could get the items you want as well as how efficiently you could get them back. Ensure that every single superfluous thing is killed from the working environment [7–9].

Fig. 2 Example of red tag

Seiri stands for filtering, dividing and dismissing (and keeping). Seiton is about planning and getting things ready. When we just concentrate on putting stuff aside, it could lose track as to why we put stuff away that is to be able to bring stuff back easily when we really need it. Seiton implies to decide when and where objects are required and also to position items in such a manner that enables productive management. The staff does not get to search for appropriate equipment or step to some other location to reach commonly used objects.

Seiton has successfully implemented at reliable tools Aligarh and has been compared in Figs. 3 and 4.

Fig. 3 Before Seito implementation

Fig. 4 After Seito implementation

Fig. 5 Seiso the workplace is looking uncluttered

2.3 *Seiso (Shine)*

It implies clean environment, which ought to be priority of everyone in the organization. Cleaning is necessary by almost everyone in organization, starting with "upper management to bottom." Cleaning should be done not merely for the purpose of sanitation, however, for a reason. To maintain a healthy portrait of neatness, everyone should be personally concerned for cleanup. The workforce must be assigned zone-wise assignments. Regular cleaning is indeed useful for detecting equipment failure. In order to work efficiently, a healthy, tidy and tidy area provides focus.

- Determine Shine targets—To make target-based shine procedure regularly as per the sorted items of the red Tag
- Storage space, equipment or empty space
- Determine shine assignments
- Divide work reliant on the zone of cleaning
- Start to shine
- Clean completely.

Implementation of Seiso is shown as in Fig. 5.

2.4 *Seiketsu (Standardizing)*

It basically makes the very first 3s a common procedure by adding simple practise for "Sorting, Straightening and Scrubbing." Daily audits of 5s should be performed with rankings which should be shown for every S.

It is important to promote displaying via images. The attention will be on effective communication and standardization for 5s. Seiketsu is applied at reliable tools Aligarh. Reliable tools effectively actualized Seiketsu. Figures 6 and 7 show the course of action of records before execution of Seiketsu and after the usage. In the

Fig. 6 Condition of workplace before Seiketsu

Fig. 7 Condition of workplace after Seiteksu

primary course of action, there are potential outcomes of missing documents. In any case, in the second case, it is a lot simpler not to turn out badly while setting and recovering the records.

2.5 Shitsuke (Sustain)

In the sustain, it implies supporting, networking and training to guarantee that it is a component of the organization. It is appropriate to continue to practice four behaviors till they become natural. This may involve appointing a group to really be accountable to monitoring the implementation of the 5s. The ideals of 5s are for all. If executives do not obey, it is pointless to ask subordinates to adopt 5s as well. This mechanism encourages individuals to be organized. Everybody in the working environment should treat it they would their own home.

- Occasional office the board association is required to watch that the initial four S's are actualized splendidly.
- Representatives must make it a piece of their everyday work and not an activity constrained upon them.
- Commitment, duty, dedication and genuineness are required in usage of 5S on regular routine.
- Senior the executives should start a festival for the absolute 5S usage and be a functioning part in the all-out procedure in starting and conveying forward the program.

- Senior the executives ought to do an occasional survey of the status of 5S.
- Assessments of initial three S's ought to be done and the outcomes showed on 5S board routinely.
- Single point exercises ought to be utilized to impart the norms for how 5S work ought to be finished.
- Proprietors lead 5S Kaizen exercises and archive results. Proprietors (administrators) complete everyday check sheets to control factors that quicken weakening of hardware and to keep clean working environments that help construct pride.

3 Conclusions

Constant overhauls have gotten especially noteworthy in the Indian circumstance in the earlier decade. The reason for this is the low availability of records, and the need to achieve overall quality measures inside the open resources. To achieve these, various associations are grasping the methodology of lean manufacturing.

The 5S the structure is an average starting stage for all improvement tries meaning to drive out waste from the collecting technique and in the long run improving an association's primary concern age by improving things and organizations and cutting down expenses. The final product of a 5S execution is a critical decrease in space required for existing activities. Laborers improve their workspaces by cleaning and sorting out them. Devices and materials are named and put away in composed stockpiling areas. Racking and racks advance the capacity of things in a littler impression, improving the request picking process by taking out the need to look for things. Once completely actualized, the 5S framework raises profitability, makes positive brand encounters and improves proficiency and association.

Implementation of 5s has successfully done in that industry keeping the point of view of its sustenance in future by giving proper training to the employs laborers. Not exclusively will representatives feel good about where they work, yet the consequences for nonstop improvement will add to less duplication, better quality and shorter lead times. 5S is not just a housekeeping technique; it is an imaginative answer for expanding productivity. 5S is an entire culture that expands generation, improves quality, lessens costs, conveys on schedule, improves wellbeing and improves profound quality.

Future road map for this implementation would be to ensure that all the pillars of are properly defined and sustained throughout the project and afterward.

References

1. Agrahari R (2015) Implementation of 5S methodology in the small scale industry: a case study. Int J Sci Technol Res 4(4):130–137
2. Bahadorpoor Z (2018) Implementation of 5S methodology in public libraries: readiness assessment library philosophy and practice, pp 1–16

3. Gapp R (2008) Implementing 5S within a Japanese context: an integrated management system. Manage Decis 46(4):565–579
4. Haslinda M, Muliati S (2018) Implementation of 5S in manufacturing industry: a case of foreign workers in Melaka. In: MATEC WEB OF CONFERENCES 150, 05034 MATEC Web of Conferences, EDP Sciences
5. Kapur D (2018) Food safety practices and 5s implementation in storage area of foods industry: a case study. Int J Sci Res Sci Technol 4(2):1019–1039
6. Patel VC, Thakkar H (2014) A case study: 5s implementation in ceramics manufacturing company. Bonfring Int J Ind Eng Manage Sci 4(3):132–139
7. Patel VC, Thakkar H (2014) Review on implementation of 5S in various organization Int J Eng Res Appl 4(3):774–779
8. Sorooshian S (2012) Case report: experience of 5S implementation. J Appl Sci Res 8(7):3855–3859
9. Wojtynek L(2018) Implementation of lean 5s methodology in logistic enterprise. Res Log Prod 8(2):179–187
10. Islam A, Dwivedi V (2020) Application of statistical and management methodologies in a small-scale industry and there after effects. In: International conference on engineering, technology and management for the sustainable development, vol 83, pp 23180–23190
11. Red Tag | Lean term from the Continuous Improvement Companion. https://www.velaction.com/red-tag/

Availability Analysis of Shoe Manufacturing Unit Using Petri Nets

Ankur Bahl, Satnam Singh, and Jaiinderpreet Singh

Abstract Today industries are under immense pressure to achieve production targets and high payback ratios for survival. This can only be accomplished by assuring the highest levels of system availability. Availability analysis is helpful to accomplish minimum failures objective or to enhance the mean time between failures to meet the production targets. The paper proposes Petri net-based modelling of a shoe manufacturing unit to compute the availability of various machines of the unit. The industrial system consists of four machines/subsystems namely sewing machine, skiving machine, moulding machine, and toe puffing machine. The dynamic nature of the system is studied through the MC simulation. The effect of failure and repair times on system availability have been analysed and discussed. The results can be presented to the maintenance personnel which; in turn; will help them to study the dynamic behavior of the plant and to decide maintenance priorities.

Keywords Petri nets · Availability · Simulation · Maintenance

1 Introduction

In the present era of modern technology, it is required to build highly complex systems to achieve production targets. To achieve production targets, it is desired by the plant managers that these systems should not fail. But during the operation of the plant, the failure of subsystems may occur due to many reasons such as wear and tear of the equipment or some faults related to sensing devices. Also, faulty operating procedures adopted by equipment operators may sometimes lead to unavoidable failures. Failure is an unavoidable phenomenon. Hence it is desired to increase the system availability. Availability of the system can be enhanced either by redundancy or by increasing the reliability of the components. The increase in availability is achieved if the components are more reliable and more effective maintenance measures are being adopted. As availability is a function of both reliability and maintainability so

A. Bahl (✉) · S. Singh · J. Singh
School of Mechanical Engineering, Lovely Professional University, Phagwara, Punjab, India
e-mail: ankur.bahl@lpu.co.in

A. K. Dubey et al. (eds.), *Recent Trends in Industrial and Production Engineering*,
Lecture Notes in Mechanical Engineering,
https://doi.org/10.1007/978-981-16-3135-1_5

they are used to assess the plant performance. In broader terms, the availability of the system is measured as the proportion of uptime to the total time the system is in service. Various tools and techniques have been discussed by various authors for the availability analysis of complex systems in the literature. Zhang et al. [1] used six sigma and Gauss–Legendre quadrature formula for carrying out the reliability analysis of a complex system. Byun et al. [2] discussed a matrix-based system reliability method to identify the intricate dependence between the various component failure for reliability analysis.

Few researchers [3] discussed the RAM analysis of tunnel boring machines using Markov chains. The effect of failure and rates of various components of the machine on the RAM analysis was analysed. Agrawal et al. [4] carried the reliability analysis of the dragline using FMEA and Bathtub curve. Pandey et al. [5] presented the reliability analysis of the butter oil processing unit using Markov chains considering the constant FRRs of the components. Garg et al. [6] used a Supplementary Variable Technique for carrying out the long-run steady-state availability analysis of the yarn plant. Khorshidi et al. [7] studied the reliability analysis of complex large systems using the FMEA technique. Murthy et al. [8] performed the reliability analysis using a Markov model of the phase measurement unit considering the transient system states. Imakhlaf et al. [9] discussed the reliability analysis of the non-coherent systems using binary decision diagrams. Choi et al. [10] proposed the fault tree model to analyse the reliability and availability of seabed storage tanks. A four-step procedure of data collection and modelling the system for the reliability assessment was proposed. Modgil et al. [11] used Markov-based approach to carry out performance modelling of a shoe manufacturing unit. Ahmed et al. [12] presented the availability analysis of gas sweetening plant. Though the fault tree, decision tree, and reliability block diagrams have been used by various researchers for the availability analysis, but it is difficult to form them for large complex systems.

Many researchers have used the Markov chains and Supplementary Variable Technique for the reliability analysis of the complex systems. These techniques are helpful in calculation of long-run availability but they are difficult to formulate. Moreover, they involve complex differential equations to calculate the availability of the system which needs more computational efforts. It is hard to produce state transition diagrams for complex systems. These traditional tools and techniques do not address problems like complex parallelism, process dependency, and resource constraints. These problems are well addressed by Petri nets; which is an effective modelling tool to represent complex systems. Petri net is a graphical tool; is used as communication visual aid and being a mathematical tool; helps to study the dynamic behaviour of the system by setting up the governing equation to the model.

Many authors have discussed the various applications of Petri nets in different fields. Bahl et al. [13] discussed the use of Petri nets in the availability analysis of a distillery plant. Sachdeva et al. [14] presented the application of stochastic Petri nets to study the behaviour of the feeding system of a paper plant. Patel and Joshi [15] carried out the analysis of manufacturing systems using Petri nets. Liu et al. [16] presented the use of DSPN for the performance analysis of a subsea low preventer unit. Kaakai et al. [17] presented the application of hybrid Petri nets in the design of new stations for the transit of passengers which help the authorities about the safety and security parameters. Khan et al. [18] discussed the applications of Petri nets in deciding the control strategies to reduce the railway. Fecarotti et al. [19] carried out the performance modelling of fuel cell systems using stochastic Petri nets.

Inspired by the great modelling work done by various researchers; an attempt has been made in this paper; to use the Petri net as a modelling tool to study the dynamic behaviour of the shoe manufacturing unit. The detailed analysis of the effect of failure and repair times of various units of the shoe manufacturing unit on the overall system availability has been performed.

2 Petri Nets

Algebraically the Petri nets are represented by 5- tuple

$$PN = \{P, T, A, W, M_0\}$$

P is the finite set of places $\{P_1, P_2, \ldots, P_n\}$.
T is the finite set of transitions $\{T_1, T_2, \ldots, T_n\}$.
A is a set of directed arcs.
W is a weight function that takes values 1, 2, 3, ...
M_0 is the initial marking.

A marking M can be represented by vector M = {m1, m2, ..., mi} where mi represents the number of tokens in the place Pi.

Whereas graphically the PN are represented by the circles representing the places, rectangles representing the transitions, arrows representing the arc. There are two arcs input arcs (places to transition) and output arc (transition to places). The dynamic behaviour of the system is depicted by the movement of the token from place to place. The elements of the Petri net are represented in Table 1.

3 System Description

The shoe manufacturing plant has four different machines, namely sewing machine, skiving machine, moulding machine, and toe puffing machine. The process of shoe manufacturing starts with input raw material in the form of leather and polymer sheets of various sizes and shapes are which are fed to the sewing and skiving machine. In a sewing machine, the leather sheets of different shapes and sizes are stitched together. In a skiving machine, the polymer sheets are trimmed to the desired shape and size. Both the leather and polymer sheets are then glued together manually by skilled workers and sent to the moulding machine where the heat is provided to them and pressure is applied to make a strong bond between them. After this tip of the shoe is attached to the shoe upper mould by the toe upper puffing machine. The final product

Table 1 Elements of Petri net

PLACE	Place represents the conditions of an event
TRANSITION	Transitions represent the event of a system. Its firing leads to change in the state of a system
IMMEDIATE TRANSITION	This transition is associated with zero delays
ARC	Arc represents the relation between place and transition
TOKEN	Tokens used to represent the current state of the system. Place holding token represents the working state of the system

is sent for quality check and ready for dispatch. There is one unit of each sewing and skiving machine. The failure of any of these machines leads to failure of the system and production comes to halt. There are two units of each moulding machine and toe puffing machine. The failure of any one unit of these machines reduces the plant capacity but does not brings the whole system to halt. The schematic diagram of the system is shown in Fig. 1.

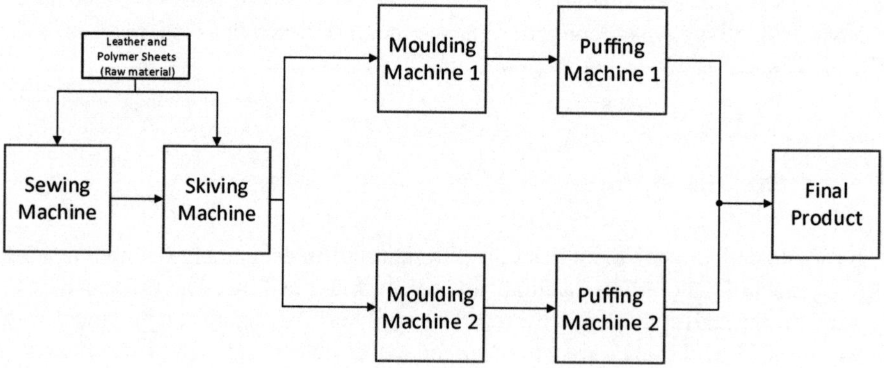

Fig. 1 Schematic flow diagram of a shoe manufacturing plant

3.1 Assumptions for the Modelling

1. A crew of Six repairmen is available.
2. The repair of fault is addressed without delay.
3. Concurrent failure can happen in the system.
4. Failure and repair rates are independent of each other
5. The priority policy of FCFS is considered.
6. After repair, the system state is restored to the original.

4 Petri Net-Based System Modelling

To carry out the performance analysis of a shoe manufacturing plant, the interactions among various units of the system are considered. Petri nets are applied to model the interaction between them. If the failures occur simultaneously in the various units of the systems, then repair is done on the first come first serve basis and it also depends upon several repair facilities available at the time of failure. Figure 2 shows the Petri net model representing the above system. Table 2 represents the description of the Petri net model.

4.1 Representation of Dynamic Behaviour of the System

Figure 2 shows the butter oil processing system in the upstate. The presence of tokens in places 1, 4, 7, 11, 13, and 16 make the transitions Tr1, Tr4, Tr7, Tr10, Tr13, Tr16 enabled. As soon as the stochastic delay of transition Tr1 is achieved the transition fires lead to movement of a token from place 1 to place 2. Place 2 representing the waiting for the repair condition of the unit. The token in place 2 enables the immediate transition which moves the token to place 3 which represents the unit is under repair. The transition Tr2 gets enabled with the presence of token in place 3. As the stochastic delay becomes equal to repair rate the transition Tr2 fires which put the token from place 3 to place 1. Similarly, all the units will behave, and depending upon the different guard conditions the tokens from Place System_up_State move to System_down_state. When anyone unit of a clarifier or the filling unit fails the system will run at reduced capacity which is represented by a token in system_reduced_capacity place.

4.2 Performance Analysis of the System

The performance analysis of the shoe manufacturing plant is done using the MC simulation of the PN model using the GRIF2020 Petri net module. The simulation evaluates the plant availability. It runs for 10,000 replications for 1 year at a 95% confidence level. The system performance analysis is evaluated by considering the effect of variation of Failure and repair rates of the various units the plant availability. Also, the system is evaluated in terms of the plant working at reduced capacity. The data-related failure and repair times of all units of the system are shown in Table 3.

The effect of variation of failure and repair times on the availability of the plant is shown in Table 4. As shown in a table with an increase in failure times and a decrease in repair times of various machines the system availability increases by 5%, 9.5%, 2.3%, and 2% for a sewing machine, skiving machine, moulding machine, and toe

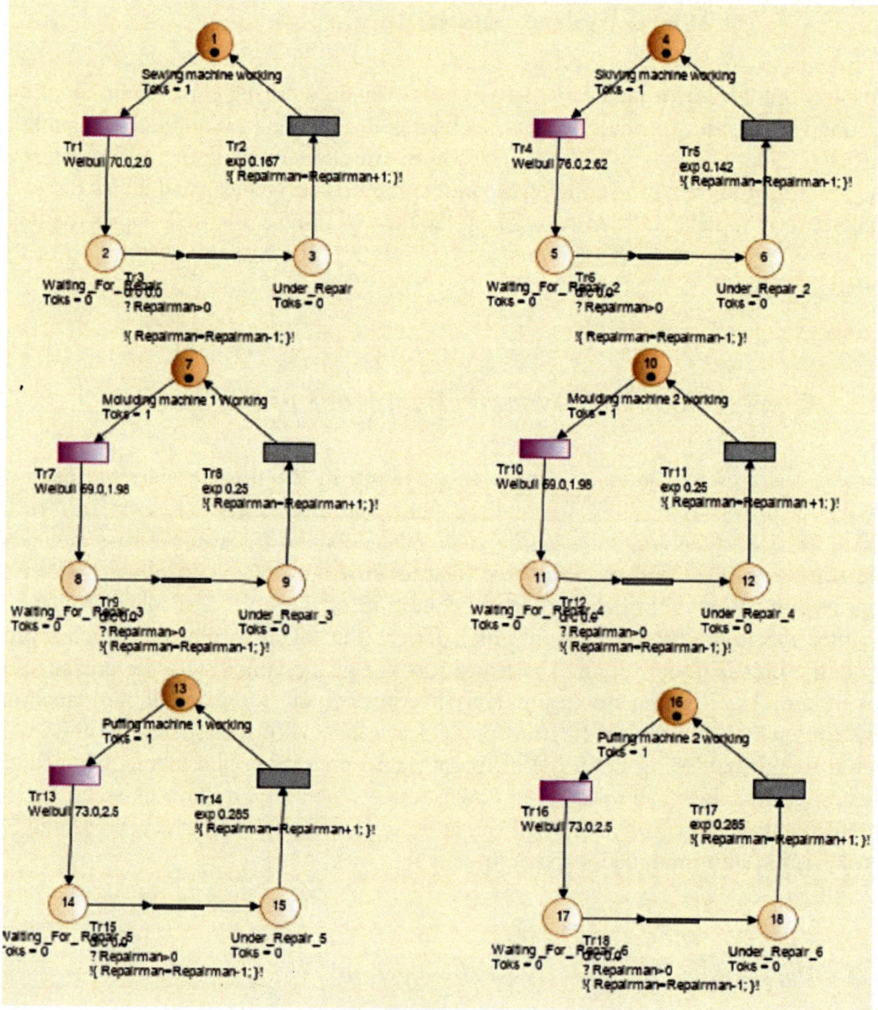

Fig. 2 Petri net model of shoe manufacturing plant

Fig. 2 (continued)

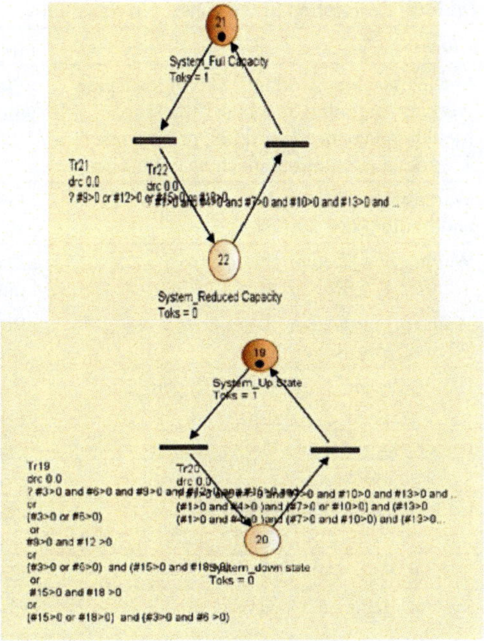

puffing machine respectively. This shows that the impact of the failure of the skiving machine on the system is highest as compared to other machines. Therefore, plant engineers need to pay more attention to it.

5 Conclusion

The Petri nets have been observed as an influential tool to study and analyse the dynamic behaviour of complex systems. In this paper, the Petri net-based model for the shoe manufacturing unit has been developed to find the system availability. The PN modelling helps the decision-makers to understand the interaction about the various units of the system. The Weibull and exponential distributions are considered for failure and repair times of all the four machines. Upon critical examination of the analysis of the failure and repair times, it is revealed that the skiving machine has a significant impact on system availability. However, other machines such as moulding and toe puffing machines have a lesser impact on system availability. The performance analysis helps the plant managers to pay attention to the critical subunits of systems. Also, it helps them make decisions related to the maintenance priority and spare part inventory management for achieving the long-term availability of the plant. Apart from these recompenses, the performance analysis can also be used by plant managers for exploring cost–benefit analysis and replacement-related decisions.

Table 2 Description of the places and transitions

Places	*Description*
Sewing machine working, Skiving machine working, moulding machine1 working, moulding machine2 working, puffing machine 1 working and puffing machine 2 working	Places represent the working states of various machines
Waiting_For_Repair_1 andWaiting_For_Repair_2	Places represent the sewing and the skiving machine has failed and is waiting for repair
Waiting_For_Repair_3 and Waiting_For_Repair_4	Places represent the moulding machine has failed and is waiting for repair
Waiting_For_Repair_5 and Waiting_For_Repair_6	Places represent the puffing machine has failed and is waiting for repair
Under_Repair_1 and Under_Repair_2	Places represent the sewing and skiving machine has gone under repair
Under_Repair_3 and Under_Repair_4	Places represent the moulding machine has gone under repair
Under_Repair_5 and Under_Repair_6	Places represent the puffing machine has gone under repair
System_Full_ Capacity	Place represents that all the units are working
System_Reduced_Capacity	Place represents that some of the units are not in working and the system is running with reduced capacity
System_Up_State	Place represent that the system is in a working state
System_Down_State	Place represents that the system is in the downstate or the system is not available
Transition	*Description*
Tr1, Tr4, Tr7, Tr10, Tr13, Tr16	Represents the timed transitions associated with stochastic delay all the units and firing of these transitions lead to failure of the corresponding unit. The scholastic delays are related to the failure rates of the units
Tr2, Tr5, Tr8, Tr11, Tr14, Tr17	Represents the timed transitions associated with stochastic delay all the units and firing of these transitions lead to repair of the corresponding unit. The scholastic delays are related to repair rates of the units

Table 3 Failure and repair data for various units of shoe manufacturing plant

Machines	Failure data (Weibull distribution)			Repair data (Exponential distribution)
	Scale factor (θ)	Shape factor(β)	MTBF (h)	MTTR (h)
Sewing machine	70	2.0	63	6
Skiving machine	76	2.62	74	7
Moulding machine	69	1.98	65	4
Puffing machine	73	2.5	79	3.5

Table 4 Effect of failure and repair times on the system availability

Units	Variation of failure times (repair times)	System availability
Sewing machine	55–90	0.90–0.86
	(7–4)	(0.7–0.84)
Skiving machine	60–90	0.92–0.84
	(9–6)	(0.81–0.89)
Moulding machine	45–80	0.88–0.86
	(5–3)	(0.86–0.88)
Puffing machine	50–85	0.87–0.85
	(4–1)	(0.87–0.89)

References

1. Zhang X, Gao H, Huang HZ, Li YF, Mi J (2018) Dynamic reliability modeling for system analysis under complex load. Reliab Eng Syst Saf 180:345–351
2. Byun JE, Noh HM, Song J (2017) Reliability growth analysis of k-out-of-N systems using matrix-based system reliability method. Reliab Eng Syst Saf 165:410–421
3. Agrawal AK, Murthy VM, Chattopadhyaya S (2019) Investigations into reliability, maintainability and availability of tunnel boring machine operating in mixed ground condition using Markov chains. Eng Fail Anal 105:477–489
4. Pandey P, Mukhopadhyay AK, Chattopadhyaya S (2018) Reliability analysis and failure rate evaluation for critical subsystems of the dragline. J Brazilian Soc Mech Sci Eng 40(2):50
5. Gupta P, Lal AK, Sharma RK, Singh J (2005) Numerical analysis of reliability and availability of the serial processes in butter-oil processing plant. Int J Qual Reliab Manage 22(3):303–316
6. Garg S, Singh J, Singh DV (2010) Mathematical modeling and performance analysis of combed yarn production system: based on few data. Appl Math Model 34(11):3300–3308
7. Khorshidi HA, Gunawan I, Ibrahim MY (2016) Data-driven system reliability and failure behavior modeling using FMECA. IEEE Trans Ind Inf 12(3):1253–1260
8. Murthy C, Mishra A, Ghosh D, Roy DS, Mohanta DK (2014) Reliability analysis of phasor measurement unit using hidden Markov model. IEEE Syst J 8(4):1293–1301
9. Imakhlaf AJ, Hou Y, Sallak M (2017) Evaluation of the reliability of non-coherent systems using Binary Decision Diagrams. IFAC-PapersOnLine 50(1):12243–12248
10. Choi IH, Chang D (2016) Reliability and availability assessment of seabed storage tanks using fault tree analysis. Ocean Eng 120:1–4

11. Modgil V, Sharma SK, Singh J (2013) Performance modeling and availability analysis of shoe upper manufacturing unit. Int J Qual Reliab Manage 30(8):816–831
12. Ahmed Q, Khan FI, Raza SA (2014) A risk-based availability estimation using Markov method. Int J Qual Reliab Manage 31(2):106–128
13. Bahl A, Sachdeva A, Garg RK (2018) Availability analysis of distillery plant using petri nets. Int J Qual Reliab Manage 35(10):2373–2387
14. Sachdeva A, Kumar P, Kumar D (2009) Behavioral and performance analysis of feeding system using stochastic reward nets. Int J Adv Manuf Technol 45(1–2):156–169
15. Patel AM, Joshi AY (2013) Modeling and analysis of a manufacturing system with deadlocks to generate the reachability tree using petri net system. Procedia Eng 64:775–784
16. Liu Z, Liu Y, Cai B, Li X, Tian X (2015) Application of Petri nets to performance evaluation of subsea blowout preventer system. ISA Trans 54:240
17. Kaakai F, Hayat S, El Moudni A (2006) Simulation of railway stations based on hybrid petri nets. Anal Des Hybrid Syst 39(5):50–55
18. Khan SA, Zafar NA, Ahmad F, Islam S (2014) Extending Petri net to reduce control strategies of railway interlocking system. Appl Math Model 38(2):413–424
19. Fecarotti C, Andrews J, Chen R (2016) A Petri net approach for performance modelling of polymer electrolyte membrane fuel cell systems. Int J Hydrogen Energy 41(28):12242–12260

Delamination Detection and Evaluation in Composite Laminates Using Guided Ultrasonic Waves

Gurdyal Singh, Anoop Aggarwal, Sunil Kumar, and Sushil Kalra

Abstract Demand from aircraft, automobile and manufacturing industries has generated the requirement of laminated composite materials with high strength, low weight, ease in manufacturing and desired control over properties. However, composite materials are susceptible to damages such as delamination and dis-bonding caused by abrupt loading or manufacturing defects. This study explored the utilization of ultrasonic waves for delamination detection and evaluation in glass-reinforced polymer composite laminates. Simulated delamination defects of varying extent and location in between layers of composite laminates were studied with the help of ultrasonic guided waves in terms of their effects on corresponding signal signature. Relative comparison of wave signatures in healthy and damaged composite laminates helped in the evaluation of delamination location and severity. Lamb wave through transmission signatures was mapped to determine the extent of delamination and pulse-echo signatures were used to determine the location of delamination by using 0.5 MHz transducers on laminated specimens submerged in water. Results indicated a marked decrease in signal amplitude with increase in delamination size. Also, it was revealed from experiment that ultrasonic can detect the delamination in between different layers at different depths by striking ultrasonic signal at particular angle to generate particular mode for detection. This study further established the ability of ultrasonic guided waves to detect the defects in composite laminated structures occurring due to manufacturing flaws or due to work loading or degradation, thus acting as effective tool for structural health monitoring.

Keywords Delamination · Ultrasonic · Lamb waves · GFRP · Multi-layered laminates

G. Singh · A. Aggarwal (✉) · S. Kumar
Chitkara University Institute of Engineering and Technology, Chitkara University, Punjab, India
e-mail: anoop.aggarwal@chitkara.edu.in

S. Kalra
Chitkara College of Hospitality Management, Chitkara University, Punjab, India

A. K. Dubey et al. (eds.), *Recent Trends in Industrial and Production Engineering*, Lecture Notes in Mechanical Engineering,
https://doi.org/10.1007/978-981-16-3135-1_6

1 Introduction

Composite materials are being widely used for numerous applications from most advanced structures for aircraft structures to basic household appliances and sports goods. Among the various composite materials, glass fibre-reinforced polymer (GFRP) composites laminates offered good replacement for steel and other metal structure due to relative advantages such as high strength-to-weight ratio, high stiffness, corrosion resistance and desired control over properties by controlling the manufacturing/laminates [1, 2]. However, GFRP composites may develop defects arising due to different loading conditions like delamination, fibre pull-out and cracks or may develop under working condition due to environmental degradation, exposure or it may be developed due to manufacturing flaws like missing fibre and air voids which could influence the functional integrity, load-bearing capability or may lead to complete structural failure [3, 4].

Delamination defect is one of the major cause of failures in composite laminated structures [5]. Delamination is de-bonding in between the adjusted layers due to the application of load or environmental degradation. In particular, laminated composites are particularly vulnerable to delamination damage due to their weak interlaminar shear strengths and transverse tensile [6]. It is not possible to detect the damages to the composite laminates by visual inspection. Non-destructive testing (NDT) offered the reliable techniques to estimate the structural health monitoring (SHM) of engineering structures. Among the various NDT techniques, guided ultrasonic waves have proven its ability to detect hidden delamination defects effectively and efficiently with good sensitivity while inspecting large area [7].

Different researchers over the period of times have proposed different techniques for damage detection in GFRP composites. Ultrasonic guided waves offer great potential for the detection of defects in composite structures [8]. Guo et al. [9] studied the influence of delamination's on lamb waves signature and showed that reflection amplitude of the S_0 mode of lamb wave strongly dependent on the delamination depth. Staszewski et al. [10] demonstrated the ability of lamb waves amplitude profiles to capture severity and location of delamination in a composite plate by using three-dimensional (3D) laser vibrometry. Petculescu et al. [11] used ultrasonic waves to detect and determine the size, location of delaminations in unidirectional and cross-ply composites by using a group delay measurement technique and found that the delamination affects the travelling time of waves. Purekar et al. [12] showed the detection of damage caused by delamination in a composite laminate by using piezoelectric phased sensor arrays. Michaels et al. [13] studied the guided wave interactions with defects in composite materials by using guided wavefield images and frequency–wavenumber domain analysis. Yeum et al. [14] demonstrated the use of dual piezoelectric transducer network for delamination detection and concluded that interaction of anti-symmetric A_0 mode with delamination slow down their speed while it had no effect on symmetric S_0 mode. Pudipeddi et al. [15] studied the scattering and mode conversion of lamb wave during interaction with delamination using three-dimensional (3D) finite element (FE) model. However, research in this area is never enough due to tremendous potential offered by ultrasonic testing as NDT technique for health monitoring and damage detection in composite laminates.

This paper provides the further exploration for the utilization of ultrasonic guided waves for delamination detection in GFRP laminates under water-coupling mode. In this study, delaminations of varying extent were provided in between the layer of composite laminated specimen. Basic aim of this study is to map ultrasonic lamb wave signatures of specimens with seeded delamination defects of varying extent and studying the change in signal signature with delamination extent.

2 Experimental Details

2.1 Materials

Material was procured from BASF construction chemicals (India) private limited. E-Glass 900 GSM unidirectional glass fibre was used as fibrous material. MBrace base and hardener were used to prepare epoxy for composite laminates. Epoxy was prepared by mixing base and hardener in 10:4 (by weight).

2.2 Methods

Fabrication of specimens laminates, insertion of simulated delamination, experimental set-up details, measurement settings were described in the following paragraphs.

Specimens Fabrication. Glass fibre sheet of required size was cut from roll of GFRP sheet. The sheets were initially cut 50 mm more than actual sample length. The reason for overcutting was that after laminated specimen has been fabricated and cured, extra edges were trimmed in order to remove flaws after layup operation. Epoxy was prepared by mixing base and hardener in 10:4 (by weight). Approximately, 300gms of epoxy was needed to apply to both sides of the sheet of given one layer of specimen. The epoxy was applied on sheet using a steel scrapper by hand layup operation by carefully spreading it evenly on all sides of sheet. Care was taken to avoid air bubbles inclusion in epoxy while layup process. After that step another sheet was placed in the same direction (0° orientation), and once again epoxy coating was applied to create a laminated sheet. Finally, the prepared laminate specimen consists of three epoxy layers and two glass fibre sheets in between them. Then laminate sheets were left for curing under ambient conditions for seven days. Once the specimens were fully cured, they were cut to actual specimen size by using the circular saw machine.

Insertion of Simulated Delamination. Two types of delamination were put into specimens to study delamination effects. One was of varying extent and other was

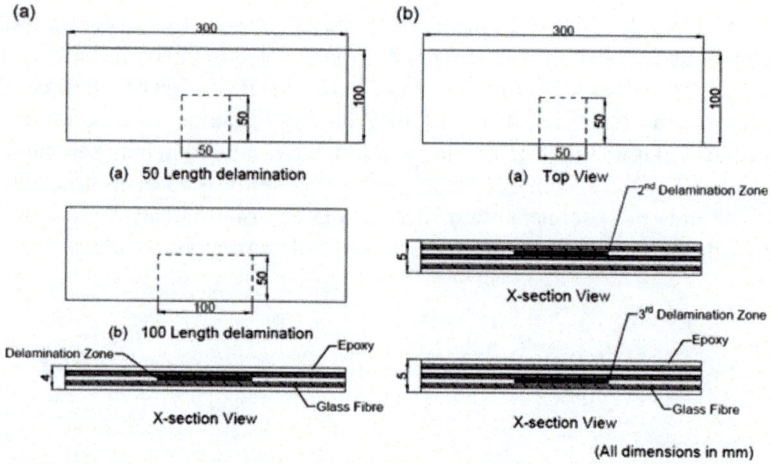

Fig. 1 Specimen with simulated delamination damage **a** varying extent, **b** varying depth

by varying depth, i.e. in between different layers. For varying extent delamination, two-layer (glass fibre) laminate was fabricated at 0° orientation and different size of delamination was created by inserting steel strips of 50 × 50 mm and 100 × 50 mm as shown in Fig. 1a. To simulate delamination at different depths, three-layer (glass fibre) unidirectional laminate was fabricated. For this, two delaminations of same size (50 × 50 mm) were created in three-layer laminated specimens at different depth locations. One delamination, called second-level delamination, was inserted in between first and second fibre layer, whereas third-level delamination of same size was induced in between second and third layers of glass fibre laminate specimens as shown in Fig. 1b.

Experimental Set-up for Measurements. The experimental set-up consists of pulser/receiver system which generates the negative spike pulse with pulse duration ranging from 10 to 70 ns. This pulse is sent to the transducer which converts the pulse to mechanical wave. Transmitting transducer is kept in contact with the composite plate using the couplant (water), and at the other end, the receiving transducer is arranged in the same way. The receiving transducer is connected with pulser/receiver system which in turn sends the received signal to computer through digitizer card. JSR Ultrasonics DPR 300 pulser/receiver system with Olympus Panametrics standard transducer of 0.5 MHz (1″ diameter) and 1 MHz (0.5″ diameter) were used. Figure 2 shows schematic representation of set-up (Fig. 2a) and actual experimental set-up (Fig. 2b). Testing was done in water as compared to air because water acts as good couplant [16]. Transducers with central frequencies of 0.5 MHz and 1 MHz were used for the experiments. As dispersion characteristics vary as function of frequency and wave mode, higher frequencies were avoided because at higher frequency, multiple wave modes exist and energy of the transducer is distributed among various modes. Multiple modes and frequencies are not considered ideal for damage detection [13]. It is also particularly difficult to excite a particular mode. Also, higher-frequency

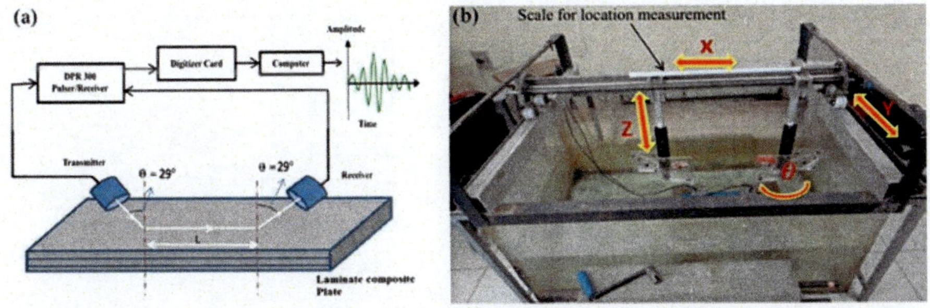

Fig. 2 **a** Schematic representation of set-up and **b** Actual experimental set-up

wave propagation has high attenuation and generally avoided for damage detection [17].

Here, we adopted the approach of finding the particular mode of lamb wave which was helpful for detecting the particular type of damage like delamination, by testing the laminate plate at various angles [18]. The angle which was particularly suitable to find particular damage more appreciably was then selected. This process was done at both frequencies, i.e. 0.5 and 1 MHz. Then angle was fixed for that damage, and signal scanning was done over laminate plates to check the delamination or health of specimens. First through transmission scanning was taken over healthy region or healthy specimen, and then compared with that of defected region or specimen with simulated delamination. Loss in amplitude had helped in determining the severity and location of delamination [8].

3 Results and Discussions

Measurement of delamination extent was done by generating particular mode of lamb waves. Further, exact location of delamination boundary was ascertained by using pulse-echo method. Secondly, the determination of delamination depth in between different layers was presented.

3.1 Delamination of Varying Extent (Length)

Delamination extent was evaluated on specimen having different length delamination in two-layer composite laminates. Transducer of 0.5 MHz frequency has been used at various angle settings to identify which angle was best suitable to excite particular mode of lamb wave that can detect delamination and its extent. After careful examination of through transmission signature at various angles, angle of 23° was found to be best suitable to find delamination in between layers of laminate. Through

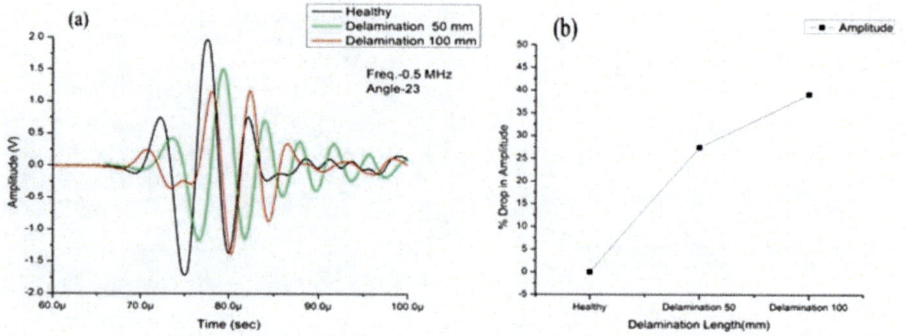

Fig. 3 **a** Through transmission signature and **b** Percentage drop of specimens having varying delamination extent

transmission signature over healthy region was taken and then compared with the signature taken over delamination region of different lengths with same angle setting and probe–probe distance as presented in Fig. 3a.

The percentage drop through signal amplitude at different delamination extent is shown in Fig. 3b. It was observed from results that with the use of right frequency-angle combination, it is possible to detect the delamination extent in laminate composite layers [19]. There was a drop of 27% in through transmission signature over delamination region having length of 50 mm as compared to signature over healthy region. When delamination length was increased to 100 mm, then percentage drop in through transmission signature was increased to 39%. This clearly indicates that with increase in delamination extent, there was drop in lamb wave signal amplitude. The region behind drop in signal amplitude may be due to the presence of multiple interfaces in delamination region [20]. This acts as hindrance in wave propagation, which results in dispersion and loss of ultrasonic signal [21]. Further, localization of delamination defect to determine its boundary was done by using pulse-echo signatures at different locations.

It is quite evident from Fig. 4 that pulse-echo signature over delamination region showed greater amplitude as compared to healthy region. This may be due to reason that incident wave was reflected back from the delamination interface instead of passing throughout thickness and getting dispersed from thickness boundary of the specimen [22]. Regions or boundaries with delamination defect were indicated by fall in amplitude of pulse transmission testing, probed by using pulse-echo technique, and exact location of defect was ascertained by reading location from the scanning set-up scale.

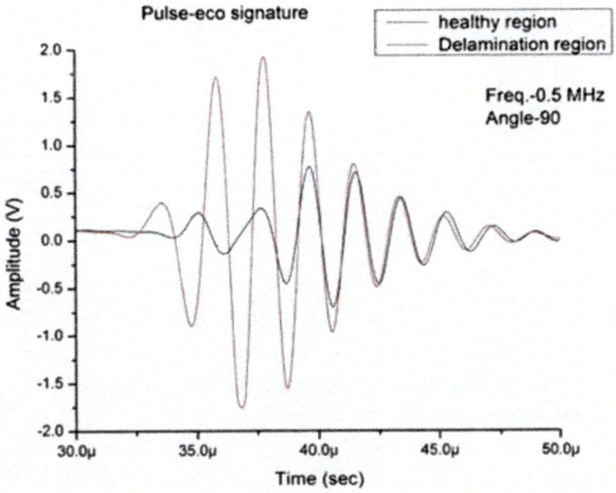

Fig. 4 Pulse-echo signature at healthy and delamination region

3.2 *Delamination Defects at Different Depths*

Lamb waves with frequency 0.5 MHz were used at different angle setting to determine the delamination at different depths of laminated specimens by generating different lamb wave modes at different angles settings for particular frequency–thickness combination. As different modes detect defects in different layers of composites, so to detect defect in particular layer that particular mode needs to be excited for defect evaluation [23]. For detection of delamination at second level, various angles were tested and it was found that angle setting of 16° generates particular mode which can detect the delamination at second level (Fig. 5a). Similarly, for detection of delamination at third level, number of angles were tested and then angle of 29° was found to be most suitable to detect the delamination at third level (Fig. 5b).

It was observed from result graphs that right frequency-angle combination is helpful in determining the delamination in between different layers of laminated composite structures by exciting particular mode of lamb wave. Here, it was observed from Fig. 6a that with 0.5 MHz probe setting at 16° angle, there was significant change

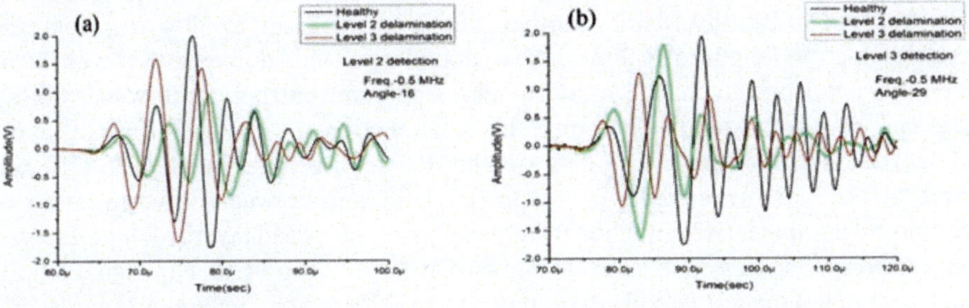

Fig. 5 Through transmission signature for **a** second-level detection, **b** third-level detection

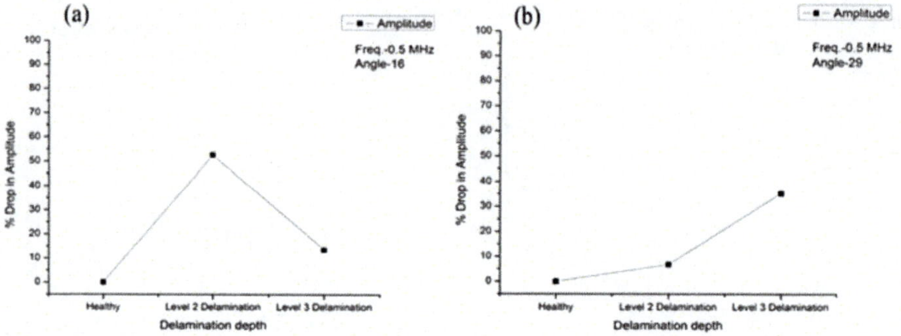

Fig. 6 Comparison of through transmission signatures for simulated delamination defects at different depths at angle **a** 16° and **b** 29°

in peak–peak amplitude wave signature between healthy zone and defective zone for specimen seeded with delamination defect at second level as compared to region having similar defect at third level. Similarly, when probe was set at angle of 29° as shown in Fig. 6b, significant change in peak–peak amplitude in lamb wave signature was observed between healthy zone and defective zone for specimen seeded with delamination defect at third level as compared to defective region with similar defect at second level. The main reason behind that as established by different researchers was different amount of energy is injected by probe into different layers of laminate at different angle setting. Maximum energy flows into particular layer at particular angle by exciting particular lamb mode. This energy of ultrasonic waves helps to determine the defect in that particular layer of composite.

4 Conclusions

Delamination study done on GFRP composite by using ultrasonic guided waves ascertain their ability to evaluate delamination severity and location in between the layers of laminated composite. From the experimentation, it was depicted that there was drop in through transmission signal (27%) over delamination length of 50 mm as compared to healthy region. Further, this drop increases to 39% with increase in delamination length to 100 mm. Also, boundary of this delamination was localized by capturing the change in pulse-echo signature. Further, lamb waves ability was successfully tested to determine the delamination in between different layers of laminated composite at different depths. By setting different angles, different lamb modes were generated which detected defects in between different layers of composite laminate. Delamination in between first and second layers of three-layered laminated specimen was detected at angle of 16° with drop of 53% in signal amplitude over delamination region as compared to healthy region. Secondly at angle 29°, delamination in between second and third layers was detected by comparing loss in amplitude (35%) over delamination region as compared to healthy region. This study further established the ability of ultrasonic guided waves to monitor the health of composite laminated structures by determining delamination defects severity and location before catastrophic failure.

References

1. Mantari JL, Oktem AS, GuedesSoares C (2011) Static and dynamic analysis of laminated composite and sandwich plates and shells by using a new higher-order shear deformation theory. Compos Struct 94:37–49
2. Quattrocchi A, Freni F, Montanini R (2019) Comparison between air-coupled ultrasonic testing and active thermography for defect identification in composite materials. Nondestr Test Eval. https://doi.org/10.1080/10589759.2019.1699084
3. Diamanti K, Soutis C (2010) Structural health monitoring techniques for aircraft composite structures. Prog Aerosp Sci 46:342–352
4. Saeedifar M, Najafabadi MA, Zarouchas D, Toudeshky HH, Jalalvand M (2018) Barely visible impact damage assessment in laminated composites using acoustic emission. Compos B Eng 152:180–192
5. Park B, An YK, Sohn H (2014) Visualization of hidden delamination and debonding in composites through noncontact laser ultrasonic scanning. Compos Sci Technol 100:10–18
6. Ip KH, Mai YW (2004) Delamination detection in smart composite beams using Lamb waves. Smart Mater Struct 13:544–551
7. Hu N, Li J, Cai Y, Yan C, Zhang Y, Qiu J, Sakai K, Liu Y, Peng X, Yan B (2012) Locating delamination in composite laminated beams using the A 0 lamb mode. Mech Adv Mater Struct 19:431–440
8. Sikdar S, Banerjee S (2016) Identification of disbond and high density core region in a honeycomb composite sandwich structure using ultrasonic guided waves. Compos Struct 152:568–578
9. Guo N, Cawley P (1993) The interaction of Lamb waves with delaminations in composite laminates. J Acoust Soc Am 94:2240–2246
10. Staszewski WJ, Mahzan S, Traynor R (2009) Health monitoring of aerospace composite structures—active and passive approach. Compos Sci Technol 69:1678–1685
11. Petculescu G, Krishnaswamy S, Achenbach JD (2008) Group delay measurements using modally selective Lamb wave transducers for detection and sizing of delaminations in composites. Smart Mater Struct 17(1):015007 (9 pp)
12. Purekar AS, Pines DJ (2010) Damage detection in thin composite laminates using piezoelectric phased sensor arrays and guided lamb wave interrogation. J Intell Mater Syst Struct 21:995–1010
13. Michaels TE, Michaels JE, Ruzzene M (2011) Frequency-wavenumber domain analysis of guided wavefields. Ultrasonics 51:452–466
14. Yeum CM, Sohn H, Ihn JB, Lim HJ (2012) Instantaneous delamination detection in a composite plate using a dual piezoelectric transducer network. Compos Struct 94:3490–3499
15. Pudipeddi GT, Ng C, Kotousov A, Manuscript D (2018) Mode conversion and scattering of Lamb waves at delaminations in composite laminates. J Water Resour Plan Manag 144:1–13
16. Kundu T, Potel C, De Belleval JF (2001) Importance of the near Lamb mode imaging of multilayered composite plates. Ultrasonics 39:283–290
17. Yang B, Xuan FZ, Chen S, Zhou S, Gao Y, Xiao B (2017) Damage localization and identification in WGF/epoxy composite laminates by using Lamb waves: Experiment and simulation. Compos Struct 165:138–147

18. Singh RK, Ramadas C, Shetty PB, Satyanarayana KG (2017) Identification of delamination interface in composite laminates using scattering characteristics of lamb wave: numerical and experimental studies. Smart Mater Struct 26(4). https://doi.org/10.1088/1361-665X/aa623c
19. Yang W, Kundu T (1998) Guided waves in multilayered plates for internal defect detection. J Eng Mech 124:311–318
20. Tian Z, Yu L, Leckey C (2015) Delamination detection and quantification on laminated composite structures with Lamb waves and wavenumber analysis. J Intell Mater Syst Struct 26:1723–1738
21. Dong J, Kim B, Locquet A, McKeon P, Declercq N, Citrin DS (2015) Nondestructive evaluation of forced delamination in glass fiber-reinforced composites by terahertz and ultrasonic waves. Compos B Eng 79:667–675
22. Infanta Mary Priya I, Vinayagam BK (2020) Detection of damages on biaxial GFRP composite material using non-destructive technique. Springer Berlin Heidelberg. https://doi.org/10.1007/s00289-020-03228-x
23. Gupta S, Rajagopal P (2018) Effect of ply orientation and through-thickness position of delamination on the reflection of fundamental symmetric S_0 Lamb mode in GFRP composite plate structures. Ultrasonics 90:109–119

Development of a Systematic Framework to Optimize the Production Process in Shop Floor Management

Varun Tripathi, Suvandan Saraswat, and Girish Dutt Gautam

Abstract In the production process, waste elimination on the shop floor is an essential task to increase production levels. Various approaches like kaizen, total quality management, six sigma, and lean six sigma are used for this purpose; however, lean management proves itself a prominent approach to achieve a high level of productivity. This information paves the way for the present study. In this article, the authors attempted to develop a framework for optimizing the production processes by identification of the wastes in shop floor management. The results of the study showed that the developed framework can reduce processing time, production time, and production costs as well as improve the quality level. The novelty of this work lies in the fact that the implementation of the developed framework has been lead to optimization and improvement in production on the shop floor. On the basis of acquired results, authors strongly believe that the present work will be highly beneficial for industry persons and researchers to improve shop floor management.

Keywords Production planning · Shop floor management · Waste elimination · Modeling · Process optimization

1 Introduction

Production systems that manufacture several products and work in highly competitive environments are focused to meet consistently high productivity levels within limited constraints [1–4]. Here, constraint means resource availability like time, machinery, cost, shop floor area, and worker. Such a competitive environment can be regulated by planning an appropriate production framework. Planning production

V. Tripathi
Accurate Institute of Management and Technology, Greater Noida, India

S. Saraswat
JSS Academy of Technical Education, Noida, India

G. D. Gautam (✉)
Mangalmay Institute of Engineering and Technology, Greater Noida, India

A. K. Dubey et al. (eds.), *Recent Trends in Industrial and Production Engineering*, Lecture Notes in Mechanical Engineering,
https://doi.org/10.1007/978-981-16-3135-1_7

framework is a critical issue in the manufacturing environment; in the worldwide industries, a very high amount is spent on production planning that requires improvement in strategy and technique [5–7]. An extensive research work has been done in previous decades on production framework development for shop floor management. However, emerging thinking that the developed production framework has not suitable for a dynamic environment where highly volatile production environments. The developed framework has been implemented for stable demand conditions.

The process improvement technique has been implemented for shop floor management extensively in previous research articles. The proposed framework has emerged as an alternative to improvement in shop floor management by using process improvement techniques. The process improvement techniques mainly include lean manufacturing, kaizen, total quality management, and six sigma [8–11]. These techniques have been implemented for productivity enhancement and resource optimization [12–14]. The modified framework of process improvement techniques maximizes productivity because it enables the identification and elimination of waste. Waste means non-productive activities of the production system.

Vinodh et al. [15] have explored shop floor improvement in a systematic manner by value stream mapping (VSM) and showed an important role of the framework in the enhancement of productivity in the manufacturing environment. Garre et al. [16] have identified problems in the production line of a pressure vessel manufacturing by implementing the lean concept. The result showed that productivity enhancement has been obtained by the elimination of non-productive activities such as excessive inventory time and transportation time. Kumar et al. [17] proposed an integrated framework for optimizing resources using VSM and fuzzy and implemented on the shop floor of a storage tanks manufacturer. The effectiveness, quality, and cost have been reduced as a result of the study by the implementation of the proposed framework. Tripathi et al. [18] proposed a framework for the elimination of non-productive activities using lean techniques. It was found in the study that the productivity of automobile industries may have improved by the implementation of the proposed framework. Yadav et al. [19] have proposed a lean six sigma (LSS) and fuzzy-based hybrid framework for quality improvement and reduction of non-productive activities. The developed framework robustness has been evaluated by obtained improvement in the manufacturing industry. Coppini et al. [20] implemented a simulation tool software using VSM in an industrial gearbox industry. It was revealed as a result that 56.7% productivity improvement, 60% reduction in lead time, 10% reduction in idle time for the supplier, and 30% reduction in idle time for foundry has been obtained by the modified framework. Salleh et al. [21] investigated total quality management and lean implementation in a forming industry using Delmia Quest Software. It was revealed as a result that the implementation of Delmia Quest Software with the proposed framework was able to obtain improvement in production. Andrade et al. [22] have implemented VSM and simulation in an assembly line of the clutch disk manufacturing industry. The result of the study revealed that the integrated approach of VSM and simulation was an efficient framework, and obtained 7% reductions in production time and also improved 10% in the use of work position. In the proposed framework in this research work, the process improvement technique has been used to improve production and optimize resource utilization.

2 Research Objective

The objective of the present research work is to investigate the effect of the process improvement technique using the proposed framework to enhance the production performance of the manufacturing unit. The following challenges and problems are considered in the present work.

1. Higher production time due to change over time (CO).
2. Higher production time due to cycle time (CT).
3. Higher production time due to non-value-added activities.
4. Higher production time due to unskilled worker.
5. Higher production time due to improper material handling.

3 Research Methodology

A methodology has been designed using production information for shop floor management by statistical analyses. The methodology is consisted of 7 steps and shown in Fig. 1.

Section 1 comprised information related to production is collected, while in Sect. 2, the production operations in the industry are shown on a map. Section 3, observed the non-productive activities in production. A mathematical model is developed in Sects. 4, 5 and applied to a mining machinery manufacturing unit, and the improvement in Sect. 6 was evaluated with the help of ANOVA, suggesting that if the model meets the criteria If applicable, then the floor management of the shop should be implemented otherwise Sect. 4 should be repeated again.

4 Research Gap

It is concluded from the literature review that most of the research was based on productivity improvement using the revised framework. In these researches, there was no study of production within resources and capital costs. It was very clear from the literature that lean manufacturing was a prominent process improvement technique and most preferred in a competitive environment, in comparison with other techniques like kaizen, total quality management, six sigma [23–25]. Lean manufacturing has a combinatorial effect on productivity improvement and it can work within limited resources [26–28]. This research work attempts to enhancement in productivity and optimization of resources within limited constraints by the lean manufacturing philosophy. Here, Constraints mean resources availability and budget.

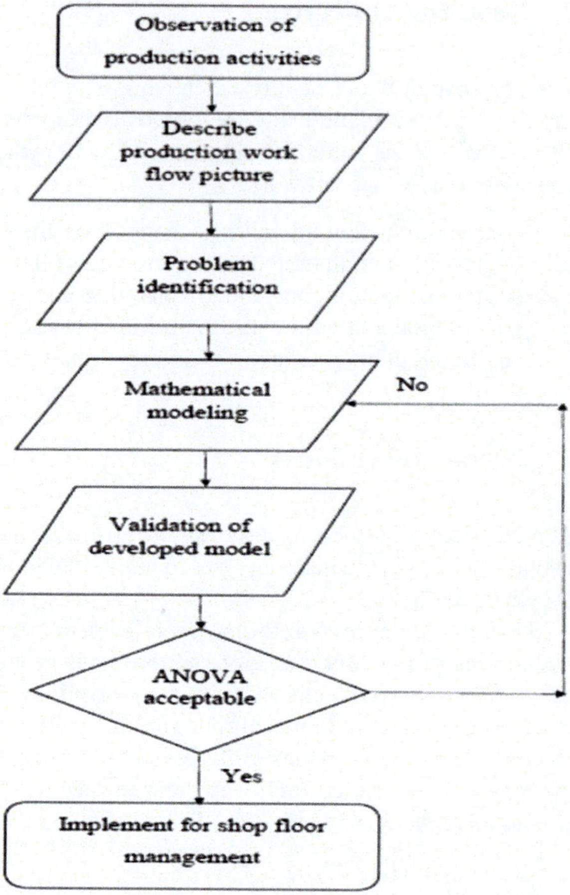

Fig. 1 Research methodology

5 Industry Description

The present research work has been performed in ABC Pvt. Ltd. manufacturing located at India. ABC Pvt. Ltd. is a leading manufacturer of earthmoving equipment (skid steer loader). Skid steer loader is a miracle in the earthmoving machinery due to its compactness and versatility. When this skid steer loader comes to the test of strength and surviving in the Indian conditions, it proves its reliability over any other earthmoving machinery. Figure 2 shows the observed production workflow on the shop floor.

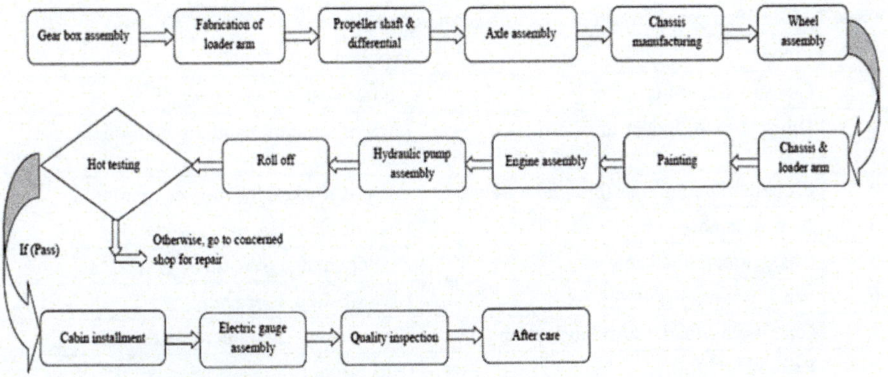

Fig. 2 Flow chart of production processes

6 Results and Discussion

It has been observed during the inspection of the shop floor that the present industry is facing problems regarding the higher production time. The problems have been identified in the changeover time, worker, cycle time, and unnecessary activities of different production processes. These problems significantly affect the budget and resources utilization of the industry. Therefore, it is required to get rid of these problems for shop floor management. Thereby in present research work, a framework has been developed using lean manufacturing to eliminate waste. Table 1 shows identified problems by the implementation of a lean manufacturing technique on the shop floor.

The developed mathematical model helps us to obtain quantities and variables of production needed for improvement in productivity. In this manner, production on the shop floor has been simulated using the Minitab software. This simulation will determine the productivity level with the optimization of resources. This mathematical model has been developed according to the following production data and shown in Table 2. Production data has been collected by shop floor observation and discussion with employees of the industry.

After analysis, an equation in terms of production has been developed. Equation (1) shows the developed second-order quadratic mathematical model.

$$
\begin{aligned}
PT = &+168.46729 + 2.47415\, CT - 0.12353 * CO \\
&- 46.32216 * NO - 62.78335 * NA + 0.067126 * CT * CO \\
&- 0.85213 * CT * NO - 0.13303 * CT * NA - 0.29786 * CO * NO \\
&- 0.45042 * CO * NA + 26.78279 * NO * NA
\end{aligned} \tag{1}
$$

An analysis of variance (ANOVA) has been performed to check the adequacy of the developed model. The results of ANOVA for *PT* are tabulated in Table 3.

Table 1 Identified problems on production shop floor

S.N	Process	Problem
1	Gearbox assembly	Higher change over time
2	Fabrication of loader arm	Lack of material handling equipment and worker
3	Propeller shaft and differential assembly	Lack of worker, location of operation
4	Axle assembly	Lack of planning
5	Chassis manufacturing	Higher change over time
6	Wheels assembly	Lack of shop floor area
7	Chassis and loader arm fabrication	Higher setup time, more workstation
8	Painting	Outsourcing
9	Engine assembly	More workstation, lack of worker
10	Hydraulic pump assembly	Lack of planning, lack of material handling equipment
11	Hydraulic motor assembly	Lack of planning, lack of material handling equipment
12	Roll-off	Unnecessary process
13	Hot testing	Lack of worker, lack in shop floor area
14	Cabin installment	Lack of planning
15	Electric gauges assembly	Lack of equipment
16	Quality inspection	Lack of planning, more workstation
17	Aftercare	Lack of planning, unnecessary process

99% confidence level is considered in performing ANOVA, which indicates that the P-value of developed models should be lower than 0.01 for adequate and reliable response model. From Table 3, it has been observed that the calculated P-value for the developed mathematical model is lower than 0.01. P-value helps to decide the rejection or failure or rejection of the null hypothesis. Therefore, it has been confirmed that the developed mathematical model has a higher degree of fitness to predict the values of the corresponding response. The calculated F-value of the response model has also been found within the acceptable range for the KW, KD, and KT as 25.47, 17.87, and 14.11 respectively.

7 Conclusions

In the present work, the authors developed a framework using lean manufacturing and validate by a case example of the earthmoving machinery manufacturing industry. This research work attempts to develop a mathematical relationship among various production parameters which include the effect of the worker, CT, CO, number of

Table 2 Production data analysis for mathematical modeling

S.N	Operation	Abbreviation of operation	Processing time (minute)	CT (minute)	CO	No. of operator (NO)	No. of activities (NA)
1	Gearbox assembly	*A*	125	90	30	2	5
2	Fabrication of loader arm	*B*	195	110	30	2	4
3	Propeller shaft & differential assembly	*C*	95	70	20	3	7
4	Axle assembly	*D*	65	35	20	3	6
5	Chassis manufacturing	*E*	210	160	40	4	5
6	Wheels assembly	*F*	85	60	35	4	3
7	Chassis and loader arm fabrication	*G*	215	175	40	5	6
8	Painting	*H*	2520	480	120	5	5
9	Engine assembly	*I*	145	90	35	3	6
10	Hydraulic pump assembly	*J*	45	30	15	2	4
11	Hydraulic motor assembly	*K*	55	35	40	2	4
12	Roll-off	*L*	45	25	45	3	2
13	Hot testing	*M*	3095	3000	55	5	7
14	Cabin installment	*N*	185	125	35	3	6
15	Electric gauges assembly	*O*	235	195	25	3	5
16	Quality inspection	*P*	125	95	20	3	4
17	Aftercare	*Q*	45	25	15	2	3

Table 3 ANOVA table of the processing time

Source	Sum of squares	Degree of freedom	Mean square	*F*-value	*P*-value
A-CT	52,861.47	1	52,861.4	159.4	0.0001
B-CO	19,473.73	1	19,473.7	58.74	0.0003
C-NO	4693.69	1	4693.69	14.16	0.0094
D-NA	389.79	1	389.79	1.18	0.3199
AB	16,215.52	1	16,215.5	48.91	0.0004
AC	5150.87	1	5150.87	15.54	0.0076
AD	387.55	1	387.55	1.17	0.3211
BC	18.41	1	18.41	0.056	0.8215
BD	344.52	1	344.52	1.04	0.3473
CD	367.05	1	2367.05	7.14	0.0369
R-Squared 99.98%			Adj R-Squared 99.96%		

steps, and process time. Based on real production information, these factors have been mathematically modeled.

The following conclusions are drawn from the present research work:

1. It has been observed that the proposed framework can improve the process performance and optimization of resources.
2. The proposed framework is enabled to reduce CT, CO, production activities, and costs.
3. It has also observed that the proposed framework remarkably reduced processing time and this reduction has been validated by mathematical modeling.
4. Processing time is improved by changing the production parameters and processing time has an effective influence on productivity, cost, and delivery time.
5. The authors of the present research work strongly believe that the present framework could be beneficial for industry persons in the enhancement of productivity level.

References

1. Monden Y (1993) Toyota production system: an integrated approach to justin time, 2nd edn. Industrial Engineering and Management, Norcross
2. Shou W, Wang J, Wu P, Wang X (2020) Lean management framework for improving maintenance operation: development and application in the oil and gas industry. Prod Plan Control 0(0):1–18
3. Ur Rehman A, Usmani YS, Umer U, Alkahtani M (2020) Lean approach to enhance manufacturing productivity: a case study of Saudi Arabian factory. Arab J Sci Eng 45(3):2263–2280

4. Tripathi V, Gautam GD, Saraswat S (2020) Process optimization methods for shop floor planning: a study. 8(10):244–247
5. Shafigh F, Defersha FM, Moussa SE (2015) A mathematical model for the design of distributed layout by considering production planning and system reconfiguration over multiple time periods. J Ind Eng Int 11(3):283–295
6. Venkat Jayanth B, Prathap P, Sivaraman P, Yogesh S, Madhu S (2020) Implementation of lean manufacturing in electronics industry. Mater Today Proc. https://doi.org/10.1016/j.matpr.2020.02.718
7. Sivaraman P, Nithyanandhan T, Lakshminarasimhan S, Manikandan S, Saifudheen M (2020) Productivity enhancement in engine assembly using lean tools and techniques. Mater Today Proc. https://doi.org/10.1016/j.matpr.2020.04.010
8. Sutharsan SM, Mohan Prasad M, Vijay S (2020) Productivity enhancement and waste management through lean philosophy in Indian manufacturing industry. Mater Today Proc. https://doi.org/10.1016/j.matpr.2020.02.976
9. Mohan Prasad M, Dhiyaneswari JM, Ridzwanul Jamaan J, Mythreyan S, Sutharsan SM (2020) A framework for lean manufacturing implementation in Indian textile industry. Mater Today Proc. https://doi.org/10.1016/j.matpr.2020.02.979
10. Balamurugan R, Kirubagharan R, Ramesh C (2020) Implementation of lean tools and techniques in a connecting rod manufacturing industry. Mater Today Proc. https://doi.org/10.1016/j.matpr.2020.03.702
11. Gopi S, Suresh A, John Sathya A (2019) Value stream mapping & Manufacturing process design for elements in an auto-ancillary unit-A case study. Mater Today Proc 22:2839–2848
12. Masuti PM, Dabade UA (2019) Lean manufacturing implementation using value stream mapping at excavator manufacturing company. Mater Today Proc 19:606–610
13. Mundra N, Mishra RP (2020) Materials today?: Proceedings impediments to lean six sigma and agile implementation?: An interpretive structural modeling. Mater Today Proc. https://doi.org/10.1016/j.matpr.2020.04.141
14. Suhardi B, Anisa N, Laksono PW (2019) Minimizing waste using lean manufacturing and ECRS principle in Indonesian furniture industry. Cogent Eng 6(1):1–13
15. Vinodh S, Arvind KR, Somanaathan M (2010) Application of value stream mapping in an Indian camshaft manufacturing organisation. J Manuf Technol Manag 21(7):888–900
16. Garre P, Nikhil Bharadwaj VVS, Shiva Shashank P, Harish M, Sai Dheeraj M (2017) Applying lean in aerospace manufacturing. Mater Today Proc 4(8):8439–8446
17. Bhuvanesh Kumar M, Parameshwaran R (2018) Fuzzy integrated QFD, FMEA framewor for the selection of lean tools in a manufacturing organisation. Prod Plan Control 29(5):403–417
18. Tripathi V, Saraswat S (2018) Lean manufacturing for shop floor of automotive industries: a study. J Exp Appl Mech 9(2):58–65
19. Yadav G, Seth D, Desai TN (2018) Application of hybrid framework to facilitate lean six sigma implementation: a manufacturing company case experience. Prod Plan Control 29(3):185–20
20. Coppini NL, Bekesas LC, Baptista EA, Vieira M, Lucato WC (2011) Value stream mapping simulation using ProModel® software. IEEE Int Conf Ind Eng Eng Manag 575–579
21. Noor NA, Kasolang S, Jaffar A (2012) Simulation of integrated total quality management (TQM) with lean manufacturing (LM) practices in forming process using Delmia Quest Procedia Eng 41(Iris):1702–1707
22. Andrade PF, Pereira VG, Del Conte EG (2016) Value stream mapping and lean simulation: a case study in automotive company. Int J Adv Manuf Technol 85(1–4):547–555
23. Garza-Reyes JA, Kumar V, Chaikittisilp S, Tan KH (2018) The effect of lean methods and tools on the environmental performance of manufacturing organisations. Int J Prod Econ 200(October 2017):170–18
24. Sahoo AK, Singh NK, Shankar R, Tiwari MK (2008) Lean philosophy: implementation in a forging company. Int J Adv Manuf Technol 36(5–6):451–46
25. Krishna Priya S, Jayakumar V, Suresh Kumar S (2020) Defect analysis and lean six sigma implementation experience in an automotive assembly line. Mater Today Proc 22:948–958

26. Ramani PV, Lingan LK (2019) Developing a lean model to reduce the design process cost of gas insulated switchgear foundation using value stream mapping - a case study. Int J Constr Manag 0(0):1–9. https://doi.org/10.1080/15623599.2019.1644756
27. Yadav G, Luthra S, Huisingh D, Mangla SK, Narkhede BE, Liu Y (2020) Development of a lean manufacturing framework to enhance its adoption within manufacturing companies in developing economies. J Clean Prod 245:118726. https://doi.org/10.1016/j.jclepro.2019.118726
28. Mahajan M, Chistopher KB, Harshan, Shiva Prasad HC (2019) Implementation of lean techniques for sustainable workflow process in Indian motor manufacturing unit. Procedia Manuf 35:1196–1204

Effect of Process Parameters on Machining of D2 Steel Using Taguchi Method

Nalin Somani, Y. K. Tyagi, and Parveen Kumar

Abstract This study investigates the effect of the EDM process parameters on machining of D2 steel by using copper electrode. In the current study, the variations in the MRR, TWR, and the SR were examined for different input process parameters. The result reveals that the input current, pulse on time as well as the pulse off time has significant affected the MRR, TWR and SR. An L9 Taguchi array was used for the designing of experimental parameters and to identify the effect of process parameters (viz. current, pulse on time and pulse off time) on the output factors {viz. MRR, TWR and SR}. Based on the S/N ratio, Optimum results for MRR and TWR were obtained for the combination for current of 20 A, T_{on} of 500 μs and T_{off} of 40 μs and the optimum results for surface roughness were found for current of 10 A, T_{on} of 400 μs and T_{off} of 40 μs. SEM was used to analyze the structural morphology of the machined surface. SEM of EDMed surface indicates that the higher pulse on duration gives rougher surface with more craters and debris than that of lower pulse on time which limits the higher pulse on time for better surface finish.

Keywords EDM · D2 steel · Taguchi method · TWR · MRR · SR

1 Introduction

EDM is a category of non-conventional machining processes. It works on the principle of thermo-electric material removal process. The thermal energy is used to generate the heat and the melting occurs due to the ionization process. It has been extensively used to produce dies and molds. During the experiment, the electrode and the workpiece remains in submerged position with dielectric fluid and the direct contact between electrode and workpiece remains absent. The work material and

N. Somani (✉) · Y. K. Tyagi
Department of Mechanical Engineering, DIT University, Dehradun 248009, India

P. Kumar
Materials and Nano Engineering Research Lab, Department of Physics, DIT University, Dehradun 248009, India

A. K. Dubey et al. (eds.), *Recent Trends in Industrial and Production Engineering*, Lecture Notes in Mechanical Engineering,
https://doi.org/10.1007/978-981-16-3135-1_8

the electrode must be electrically conductive to generate the spark [1–5]. The latter developments in the fields of EDM have progressed due to the development of new materials which are harder and difficult to machine such as D2 steel. Many researchers have reported that the machining parameters of EDM affect the MRR, TWR and the SR. Singh et al. [6] have worked on EDM and used copper and aluminum electrodes. Investigation indicates that the MRR and TWR increase with increase in current and pulse on time (T_{on}). Better results were obtained with machining from Cu electrode. Keskin et al. [7] reported that by increase in discharge duration of current during EDM process, the surface roughness was increasing. This was mainly due to increase in discharge energy during the process which expands the discharge channel. Zarepour et al. [8] have used Cu electrode for the EDMing of hot work tool steel 1.2714, which is extensively used to form the forging dies. L50 Taguchi orthogonal array was used to design the experiments. During the machining, they observed that increase in T_{on} has also increased the wear rate of tool. Similar trends were observed for the surface roughness. Guu et al. [9] have worked and analyzed the effect of process parameters on the surface texture of Fe-Mn-Al alloy. Result shows that the EDM process causes a ridged surface and induces machining damage in the surface layer, and also the surface roughness was increased. With increase in input current and T_{on}, the depth of micro-cracks and micro-voids were increased. Tai et al. [10] have worked on SKD11 tool steel and reported that the thickness of the recast layer tends to increase with incrementin the input current as well as T_{on}. Chattopadhyay et al. [11] have used EN-8 steel as a work material and EDM process was used to machine this from the Cu tool and used regression method for optimization. Result reveals that the T_{on} and the input supplied current were the most affecting the MRR and TWR. Ashtiani et al. [12] have analyzed the effect of input parameters for the machining of AISI D3 steel using different electrode. Graphite electrode was found to be more suitable as it has high MRR, precise dimension and low tool wear rate.

1.1 Taguchi Technique

It is a type of statistical method developed by G. Taguchi to improve the quality of manufactured goods in different fields [13, 14] and to design the experiment based on the input factors to opt out the suitable combination of the input process parameters and for such parameters, the output process parameters will be optimum. Taguchi is one of the robust design techniques used to improve the processing quality, to minimize the number of trials and to promote the stability in terms of quality by using the set of arrays. Taguchi methods of experimental design are fully based on well-defined orthogonal arrays. The orthogonal arrays are used to minimize the set of experiments for the particular applications [15, 16] which ultimately leads to reduce the time as well as cost. For the well-organized survey of a process, it is very important to design the set of experiments in an efficient way to gather the information. DOE is used to analyze the correlation of the responses (dependent variable) and the process variables at different levels. In DOE the quantitative formulation of responses is done.

Responses should be able to justify the final objective of the study which are the measured output of experimental work. The responses should be easily measurable. Process variables are the factors and their levels which can be used and maintained during the experiments with the desired accuracy. DOE studies the individual effect of the process variables as well as their interactive effect. Significant process variables which influence the process output can be identified. The effect of process variables on the process can also be analyzed. The experimental errors of the process are also assessed. DOE also helps to find out the optimal settings so as to achieve the desired output. Number of experiments can be decreased considerably with a well-designed experimental plan. This makes it crucial to plan a well-made set of experiments.

Analysis of variance is used to analyze the experimental values [17, 18]. In Taguchi approach, the experimental values are converted into the S/N ratio also called as signal-to-noise ratio. The "signal" shows the mean value for the *O*/*P* whereas the "noise" indicates the undesirable value for the *O*/*P*. Larger S/N ratio signifies the good performance characteristics of the experiment. So, the combination, which is having the higher S/N ration, is more preferred and treated as the optimum solution for the designed orthogonal array [19, 20].

From the literature, it is clear that the various authors have analyzed the significance of variable EDM process parameters while EDMing of different materials. However, very few literature is available on the EDMing of hardened steel for different process parameters. So, in the present study, an attempt has been made for EDMing of hardened D2 steel by varying the input parameters.

1.2 Work Materials

D2 steel was selected for the EDMing using the Cu tool. D2 steel is an air hardened and contains high chromium. It has high wear and abrasion resistant properties. The hardness of the D2 steel is in the range of 55–64 HRC which makes it difficult to machine by using the traditional machining techniques [21]. D2 steel has wide applications in shear and planer blade industries, industrial cutting tools, die and punch fabrications, etc. Heat-treated D2 steel having hardness of 62 HRC was used during the operation. Chemical composition of the hardened D2 steel determined by EDAX is shown in Table 1.

Table 1 Wt% of D2 steel

C	Cr	Co	Mn	Mo	P	S	Si	Ti	Fe
1.4	12	0.8	0.5	0.9	0.02	0.03	0.5	1.05	82.8

1.3 Electrode Material

EDM electrode should have good electrical conductivity, thermal conductivity as well as the hardness. Copper is used for various applications as it is ductile and has the high electrical conductivity, thermal conductivity, resistant to corrosion, etc. [22–24]. Based on these advantages, the pure copper (Cu) electrode was used for EDM machining of hardened D2 steel.

1.4 Experimental Details

Die sinking EDM (ZNC, EIL,3144-R50) was used for the experimental procedure. The machining was performed for 30 min. T_{on}, T_{off} and Input current were selected as the input process parameters to record the responses for MRR, TWR and the SR. Table 2 shows the experimental details. Figure 1 represents the schematic and actual machining setup of EDM.

2 Results and Discussions

The input factors for the experimentation were T_{on}, T_{off} and the input current. Experiments were designed using the L9 orthogonal array. The values were selected based on preliminary experiments. 300 to 500 μs was the range of pulse on time; 30 to 50 μs was the pulse off time and the input current was varied from 10 to 20 amp.

In order to analyze the results, S/N ratio analysis was done with MRR, TWR and surface roughness as the outputs. Minitab software was used to plot the results. The interference of various input parameter on the output parameters is shown in Table 3. Table 3 clearly indicates that the input process parameters are affecting MRR, TWR as well as SR which are validated by the S/N ratio graphs as shown in Fig. 2.

Table 2 Experimental details

Working parameters	Description
Electrode	Pure Cu electrode
Work material	D2 steel plate (50 × 50 × 10 mm^3)
Dielectric fluid	Kerosene
Current (amp), Ip	10, 15, 20
T_{on} (μs)	300, 400, 500
T_{off} (μs)	30, 40, 50
Duty cycle	1.0

2.1 Plot for S/N Ratio

For S/N ratio plots, there are three types of categories for the performance characteristics, i.e., higher-the-better, lower-the-better and nominal-the-better. In this experimental study, the higher-the-better performance characteristic is selected to obtain higher MRR, lower-the-better performance characteristic is selected to obtain minimum TWR and for SR, lower-the-better performance characteristic is selected. The experimental results for TWR, MRR and surface roughness and the corresponding S/N ratio are shown in Table 3. S/N ratio graph for MRR, TWR and SR is shown in Fig. 2a–c. From Table 3, it is clear that the maximum MRR and lowest TWR was obtained for exp. no. 9, i.e., input current of 20 A, T_{on} of 500 μs and T_{off} of 40 μs. Minimum surface roughness was obtained for exp. no. 2, i.e., input current of 10 A, T_{on} of 400 μs and T_{off} of 40 μs. From the S/N ratio main effect plots, it is

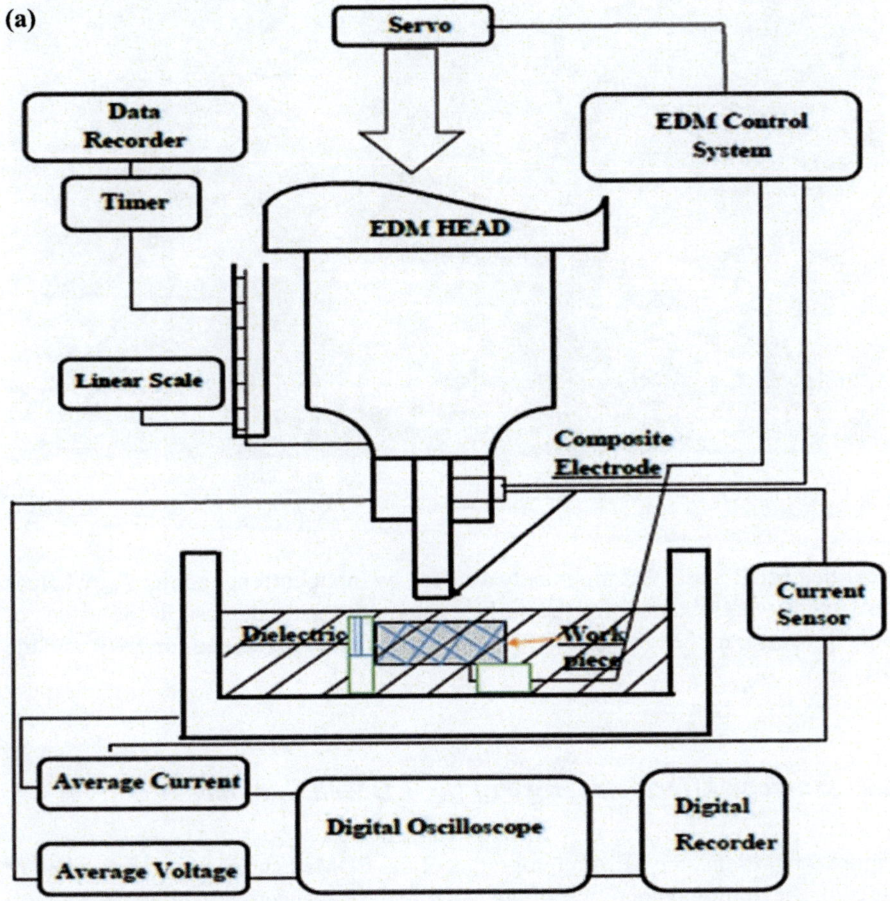

Fig. 1 **a** Schematic diagram **b** Actual setup during machining on electric discharge machine

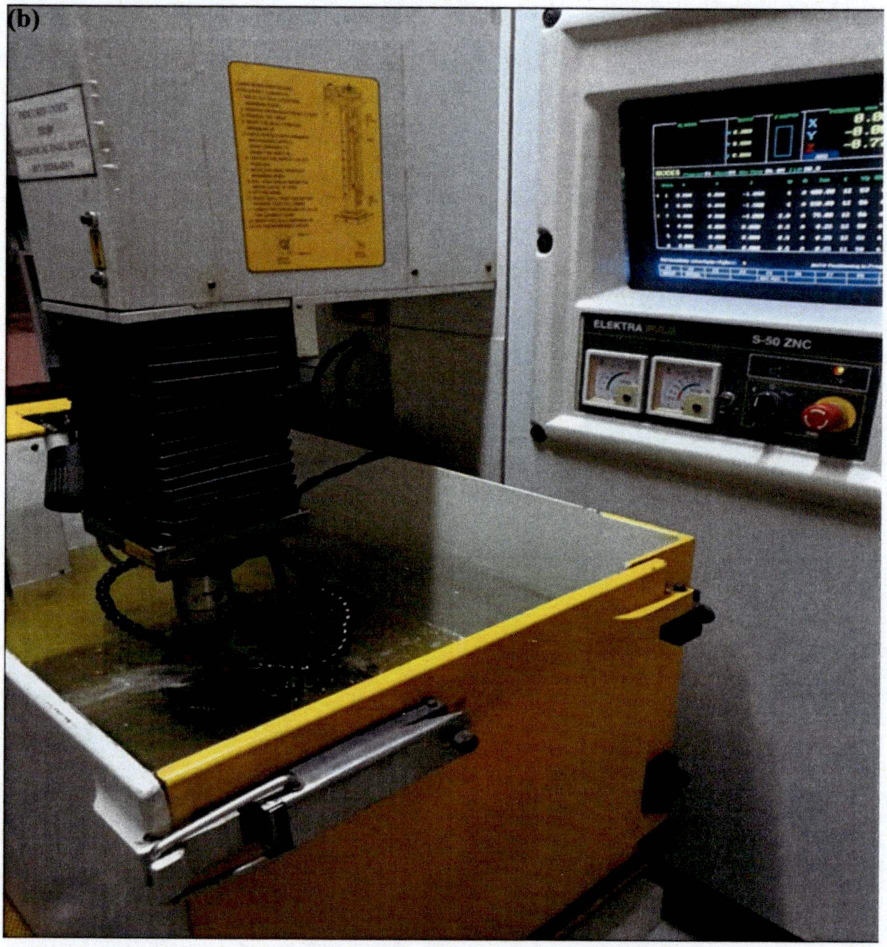

Fig. 1 (continued)

clear that MRR and TWR are mostly affected by input current and the T_{on} whereas mix trends were obtained for the surface roughness with increase in the values of input process parameters, the MRR. Reverse trends were obtained for TWR and the SR.

2.2 *Surface Micro-structure of the EDMed Surface*

Figure 3 represents the SEM micrographs of the EDMed surface machine from D2 steel for the optimum condition of the MRR and SR respectively. SEM micrograph clearly indicates that the size of craters and debris are larger for Fig. 3a due to higher

Table 3 S/N ratio of machining parameters for Cu electrode

Exp. No	Current (A)	T_{on} (μs)	T_{off} (μs)	Response values			S/N ratios		
				MRR (gm/min)	TWR (gm/min)	SR (μm)	MRR (dB)	TWR (dB)	SR (dB)
1	10	300	30	8.09	0.0099	5.31	11.490	17.741	−12.728
2	10	400	40	10.30	0.0071	5.78	20.383	14.992	−21.967
3	10	500	50	14.12	0.0046	6.57	23.525	12.245	−16.250
4	15	300	40	24.14	0.0104	6.99	26.550	16.654	−11.892
5	15	400	50	26.12	0.0091	7.25	30.214	11.245	−20.814
6	15	500	30	27.64	0.0080	7.02	30.987	9.816	−18.572
7	20	300	50	35.12	0.0054	7.58	28.114	12.381	−13.232
8	20	400	30	36.56	0.0068	7.95	32.773	8.601	−23.057
9	20	500	40	37.12	0.0065	8.02	33.764	6.976	−20.208

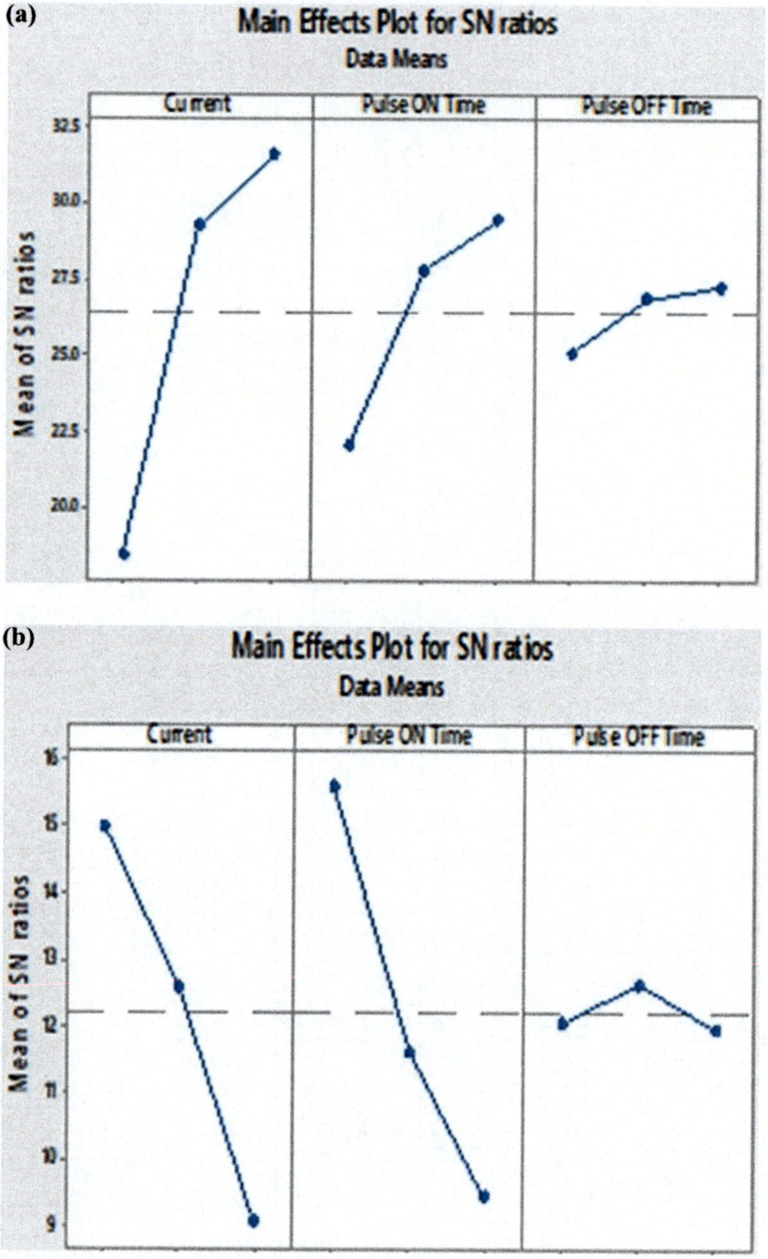

Fig. 2 S/N ratio graph for **a** MRR **b** TWR and **c** SR

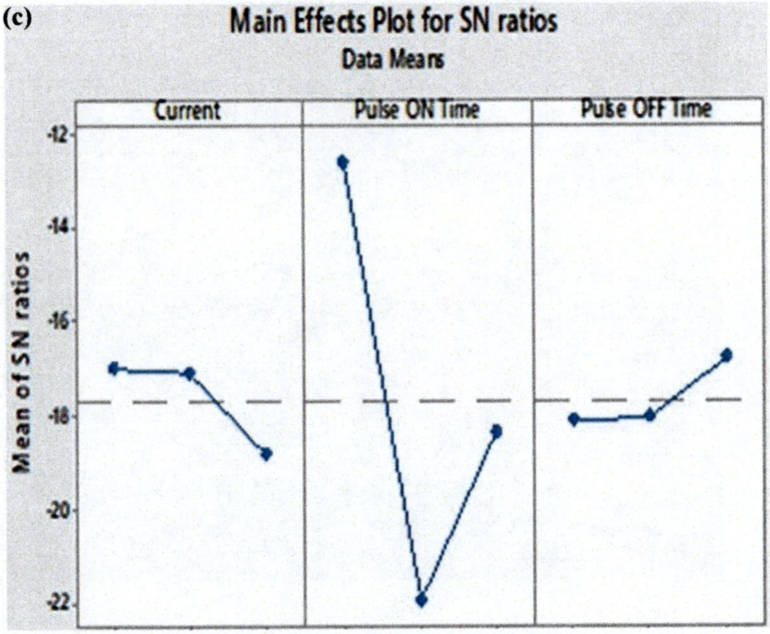

Fig. 2 (continued)

pulse on time (500 μs). This may be due to the supply of the pulse energy for longer duration. Figure 3b shows the smoother surface due to lower pulse on time (400 μs). At the smaller pulse duration, lesser amount of pulse current is supplied for small duration of time which minimizes the formation of craters and debris.

3 Conclusions

An effort has been made to saw the effect of EDM process parameters while machining D2 Steel. Machining was done by using the Cu electrode. Proper selection of input parameter plays a significant role while EDMing any material. The machining parameters taken into consideration were current, T_{on}, T_{off}. The machining characteristics were MRR, tool wear rate and SR. During the experimentation, it was found that the input current, T_{on} and T_{off} have significant affected the MRR, TWR as well as the SR. It has been noticed that the EWR increases with increment in supplied current. It has been noticed that the MRR increases with increment in supplied current and T_{on}. The optimum results for MRR and for TWR were obtained for the combination of exp. no. 9, of current, i.e., 20 A, pulse on time, i.e., 500 μs and pulse off time, i.e., 40 μs. Minimum surface roughness for the work material was obtained for exp. no. 2, i.e., input current of 10 A, T_{on} of 400 μs and

(a)

(b)

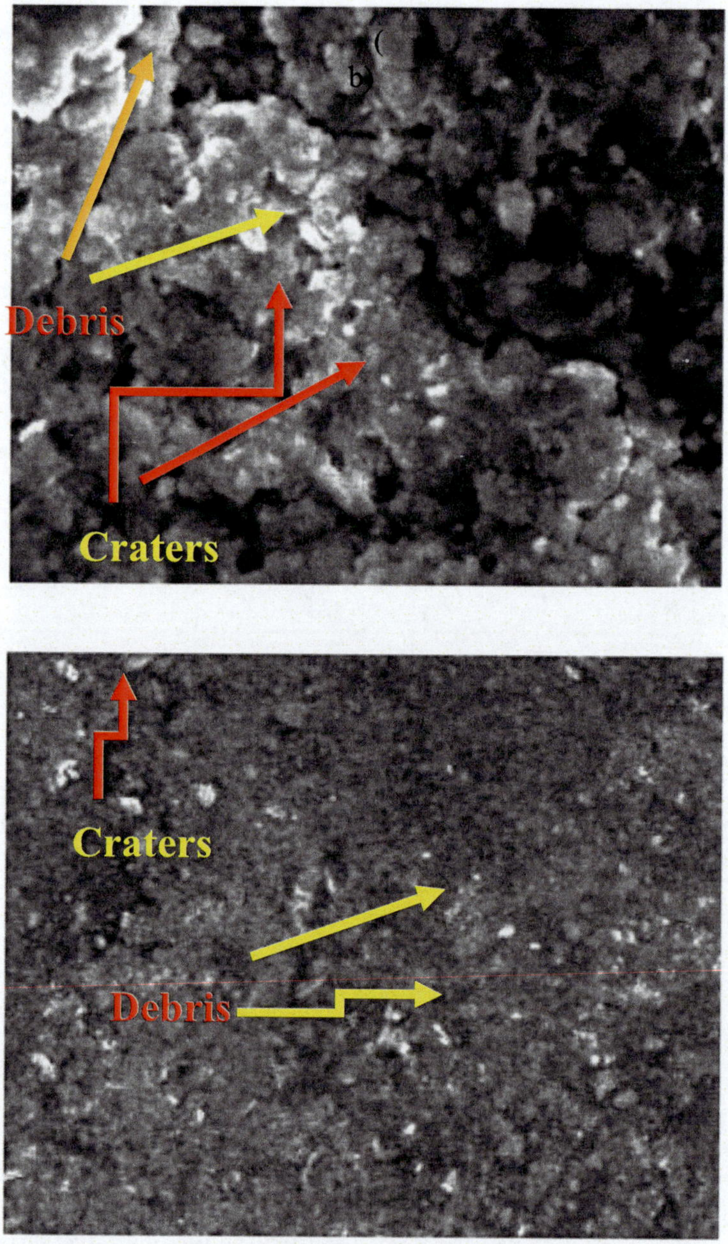

Fig. 3 Surface micrographs of the EDMed surface at T_{on} of **a** 500 μs **b** 400 μs

T_{off} of 40 μs. Surface morphology of EDMed surface indicates that the higher pulse on duration leads to increment in surface roughness which limits the higher pulse on time while EDMing of the hardened D2 steel.

References

1. Singh V, Pradhan SK (2017) Optimization of WEDM parameters using Taguchi technique and response surface methodology in machining of AISI D2 steel. Procedia Eng 97:1597–1608
2. Nipanikar SR (2004) Parameter optimization of electrode discharge machining of AISI D3 steel material by using Taguchi Method. J Eng Res Stud E-ISSN0976-7916
3. Choudhary R, Singh G (2018) Effects of process parameters on the performance of electrical discharge machining of AISI M42 high speed tool steel alloy. Mater Today Proc 56313–6320
4. Kumar NA, Abdur Rawoof HS, Viken R (2012) Investigation on the effect of process parameters for machining of EN31 (Air hardened steel) by EDM. J Eng Res Appl 2:1111–1121
5. Alekhya GVS, Indra RM (2017) Experimental investigation on electrical discharge machining of inconel 718 material. IJSRD 5:2321–2613
6. Singh S, Maheshwari S, Pandey PC (2004) some investigations into the electric discharge machining of hardened tool steel using different electrode materials. J Mater Process Technol 149:272–277
7. Keskin Y, Selcuk HH (2006) Mevlut Kizil, An experimental study for determination of the effects of machining parameters on surface roughness in electrical discharge machining. Int J Adv Manuf Technol 28:1118–1121
8. Zarepour H, Fadaei Tehrani A, Karimi D, Amini S (2007) Statistical analysis on electrode wear in EDM of tool steel DIN 1.2714 used in forging dies. J Mater Process Technol 187–188, 711–714
9. Guu YH, Ti-Kuang HM (2007) Effect of machining parameters on surface textures in EDM of Fe-Mn-Al alloy. Mater Sci Eng 466:61–67
10. Tai TY, Lu SJ (2009) Improving the fatigue life of electro-discharge machined SDK11 tool steel via the suppression of surface cracks. Int J Fatigue 31:433–438
11. Chattopadhyay KD, Verma S, Satsangi PS, Sharma PC (2009) Development of empirical model for different process parameters during rotary electrical discharge machining of copper–steel (EN-8) system". J Mater Process Technol 209:1454–1465
12. Khoshkish AG, Effects of tool electrode material on electrical discharging machining process of hardened tool AISI D3. In: Iran conference of manufacturing engineering
13. Karna SK, Sahai R (2018) An overview on Taguchi method. Int J Eng Math Sci 1:11–18
14. Athreya S, Venkatesh YD (2012) Application of Taguchi method for optimization of process parameters in improving the surface roughness of lathe facing operation. Int Refereed J Eng Sci 1:13–19
15. Backer S, Mathew C, George SK (2014) Optimization of MRR and TWR on EDM by using Taguchi's method and ANOVA. Int J Innovative Res Adv Eng 1 (2014)
16. Somani N, Gautam YK, Sharma SK, Kumar M (2018) Stastical analysis of dry sliding wear and friction behavior of Cu/SiC sintered composite. In: AIP conference proceedings, p 020018
17. Prajapati DR, Cheema DVS (2013) Optimization of weld crack expansion defect of wheel rims by using Taguchi method. Int J Innovative Res Sci Eng Technol 2
18. Gupta NK, Thakre GD, Kumar M (2019) The mechanical and tribological characteristics of Al6061 self-healing materials. Mater Res Expr 6. https://iopscience.iop.org/article/10.1088/2053-1591/ab23b1
19. Kumar S, Ghoshal SK, Arora PK (2020) Optimization of process parameters in electric discharge machining for SS420 using Taguchi technique. Adv Electromech Technol 997–1004
20. Kumar D, Mondal S (2020) Process parameters optimization of AISI M2 steel in EDM using Taguchi based TOPSIS and GRA. Mater Today Proc 26:2477–2484

21. Somani N, Kumar K, Gupta N (2020) Review on microwave cladding: a new approach. Adv Mater Proc 77–90
22. Somani N, Sharma N, Sharma A, Gautam YK, Khatri P, Solomon JAA (2018) Fabrication of Cu-SiC composite using powder metallurgy technique. Mater Today Proc 5:28136–28141
23. Somani N, Tyagi YK, Kumar P, Srivastava V, Bhowmick H (2019) Enhanced tribological properties of SiC reinforced copper metal matrix composites. Mater Res Express 6:016549
24. Gautam YK, Somani N, Kumar M, Sharma V (2018) A review on fabrication and characterization of copper metal matrix composite (CMMC). In: AIP conference proceedings, p 020017

Experimental Investigation of Water Absorptions and Charpy Test of Epoxy Composite Immersed in Different Aqueous Medium

Pankaj Singh Chandel, Nitin Kumar Gupta, Yogesh Kumar Tyagi, and Kanishka Jha

Abstract In this research work, an experimentation has been performed to investigate the water absorption and effect of rupture energy of the epoxy composite immersed in different aqueous medium. An attempt has been made to assess the rate of moisture absorption in composite materials. The percentage of moisture absorption in the specimen is observed when they are placed in four different mediums, i.e., normal water, air, half-saturated salt solution, and fully saturated salt solution. In order to check the moisture absorbed, the specimen is taken out after fixed intervals and their weights have been measured. In the beginning, a rough linear increment was observed in the weight gained by the specimens immersed in aqueous mediums. After the initial results, the results obtained were depicting non-Fickian type of moisture absorption behavior. Factors, such as initial swelling of matrix, percentage void content, edge effect, and the mechanical adhesion between fibers and matrix, greatly influenced the behavior of moisture absorption in composite. The diffusivity of the moisture in the specimens kept in air is very low owing to which a negligible change occurs in the weight of specimen immersed in air for the total period of experimentation.

Keywords Moisture absorption · E-glass fiber composite · Pressure · Fickian model · Non-Fickian behavior · Deboning · Diffusivity

---

P. S. Chandel (✉) · N. K. Gupta · Y. K. Tyagi
Department of Mechanical Engineering, DIT University, Dehradun 248001, India

N. K. Gupta
e-mail: nitin.gupta@dituniversity.edu.in

Y. K. Tyagi
e-mail: yogesh.tyagi@dituniversity.edu.in

K. Jha
Department of Mechanical Engineering, LPU, Phagwara, Punjab 144001, India
e-mail: Kanishka.21537@lpu.co.in

A. K. Dubey et al. (eds.), *Recent Trends in Industrial and Production Engineering*, Lecture Notes in Mechanical Engineering,
https://doi.org/10.1007/978-981-16-3135-1_9

1 Introduction

Composites are significantly used everywhere. We have been using composites since a long time. People used to make bricks from mud and many other materials with raw composites only. Now in present date, we make tennis racket, vehicle body and aircrafts, and many other things using composite materials only. Now a question arises what is composite? It is nothing but a multiphase material with distinct properties. Generally, they are made up of only two materials, one is the continuous phase (matrix) and the other is the discontinuous phase (reinforcement) glue to bind and reinforce fibers [1, 2].

Presently, when an engineer requires to work with materials, he/she prefers using a composite material over steel or other metals because composites are stronger, lighter, durable, much more resistant to extreme pressure and temperature, has biodegradability, and has low cost (particularly in natural fibers). These composites are used for making aircraft parts and also used in marine infrastructures, as it increases its superior strength and makes it more efficient than the older ones. The only goal of aeronautical industry to use a composite material in aircrafts is to achieve better structural properties along with lightweight, leading to improved fuel efficiency and performance from an aircraft. With the properties such as strength, lightweight, corrosion resistance, and better performance at elevated temperatures, these materials have become center of attraction in almost every field [3, 4].

With the help of composite materials, aeronautical/space engineering is upgrading day by day. In early 1970s a carbon fiber composite was introduced to engineers, which is nowadays used in making aircrafts, unlike alloy of different metals or a single metal as used earlier. High-performance composites are being increasingly used in aerospace and marine structural applications. For marine infrastructures and sustainability of materials in various moisture varying environment, glass fiber-reinforced composites are preferred better than others due to its high corrosion resistance and lightweight property, for example—in pipelines of desalination plants, turbine blades, ship building industry, water storage equipment's, boats, etc. Cost of installation is reportedly lower than that by some conventional materials like steel or aluminum [5, 6]. Hence, parts like turbine blades which are particularly immersed in water for a long period of time, moisture absorption plays a very important role, because it affects the strength of structure materials and its durability also.

Organic matrix composites are often subjected to moisture and temperature environments. These environmental effects can lead to composite degradation and consequent loss of mechanical properties. An essential feature of almost all combinations of weathering conditions is humidity. Hence, topics pertinent to the question of performance in the presence of moisture are of prime importance. These topics are commonly divided into two subjects. The first relates to factors which drive moisture into the composites, namely the penetration mechanisms. The second deals with the effects of the presence of water on the performance and durability of the composites. Several studies showed the important effects of humidity on the mechanical properties of composite materials and on their long-term behavior [7, 8]. In general, moisture diffusion in composites relies on factors such as the volume fraction of fiber, void volume, additives, humidity, and temperature. Moisture diffusion in polymer composites is seen to be governed by the variables such as humidity, temperature, and void volume. There are three different mechanisms mainly used for moisture diffusion in polymer composite. The first one is diffusion of water inside the tiny gaps, between the polymer chains. The next one is capillary action into the gaps that flows to the interfaces between the fiber and the matrix. And the last one is transportation of the microcracks in the matrix, occurring due to the swelling of fibers [9, 10].

During the research, mainly our focus was on the absorption of the moisture by the random E-glass fiber composite when they are maintained at two distinct pressures and kept in varying salt concentrated aqueous media. These media are categorized as air, normal water and two different types of salt solution according to the concentration of salt. The percentage of moisture gain is described by the weight gain at different time instants. Durability nature of the composite is predicted by the diffusion of water and its impacts on the resin properties.

2 Materials

In this research work, epoxy resin has been selected as a matrix and the random E-glass fiber were used as a reinforcement. The mixture of Araldite AY 105-Epoxy resin and HY951 Hardener were used prepare the matrix. Both the materials were mixed in the weight ratio of 10:1. All the chemical materials were procured from an Indian manufacturer Huntsman Ltd. Table 1 presents the physical properties of the Araldite AY 105 Epoxy received from the company manual. Similarly, the random E-glass has been procured from an Indian local market. Table 2 presents the physical properties of the random E-glass fiber that were used in the fabrication process.

Table 1 Properties of epoxy AY 105

Properties	Color (visual)	Specific gravity	Viscosity (Pa-s)
Araldite AY 105	Clear liquid	1:1	6–8.5

Table 2 Properties of random E-glass fiber

Properties	Density (g/cc)	Modulus (GPa)	Tensile strength (GPa)	Elongation (%)
E-glass fiber	2.57	73.5	3.5	4.8

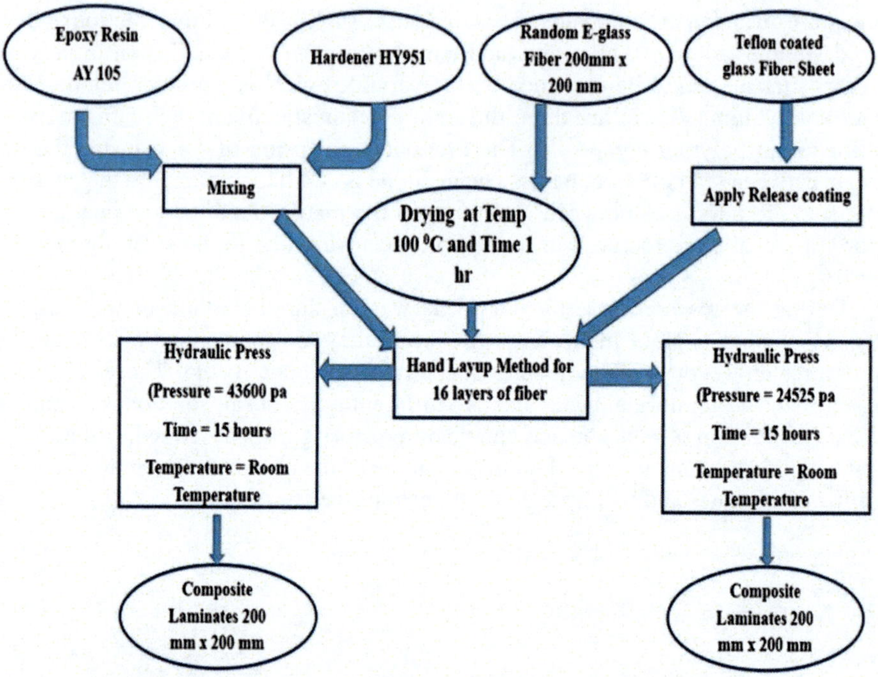

Fig. 1 Fabrication procedure of the laminated test sample with specification

2.1 *Fabrication Procedure of Test Specimens*

Two samples were made each having 16 layers of E-glass fiber composite fabricated by the conventional hand lay-up technique. The dimensions of the samples were taken as $150 \times 150 \times 4.7\ mm^3$ and $100 \times 100 \times 6.8\ 5\ mm^3$ under pressure of 43,600 Pa and 24,525 Pa, respectively, at room temperature (~20 °C) for 15 h. Figure 1 represents the fabrication procedure at various condition. And Fig. 2 represents the schematic representation of the hydraulic press.

2.2 *Procedure of Moisture Absorption Testing and Evaluation*

The water absorption tests were conducted in distilled water and based on *ASTM D57011*. For each laminate, five test specimens measuring 25×76 mm were used [11]. The differences in these pressures are to ensure the distinct percentage of void content in the corresponding samples of composite (Fig. 3).

The effect of water absorption on the fabricated sample was investigated through the ASTM D57011 procedure. Specimens are prepared with the help of two samples

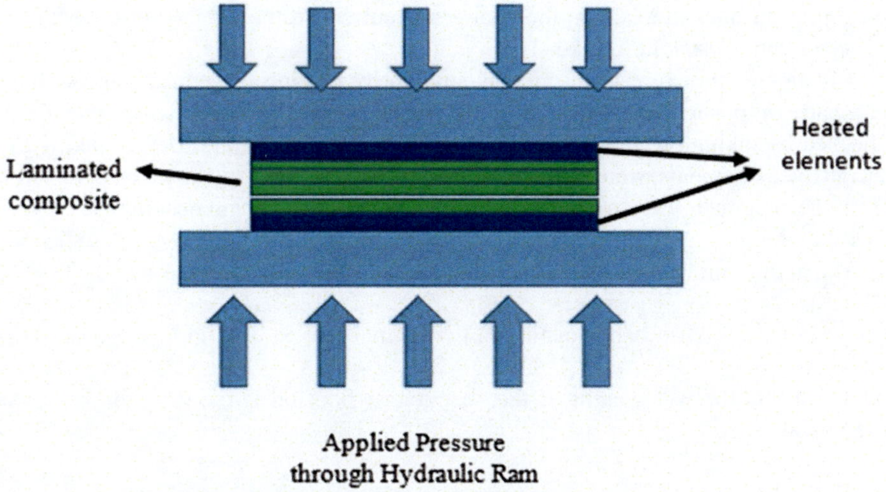

Fig. 2 Schematic representation of hydraulic press used for laminate fabrication

Fig. 3 Actual fabricated sample used for the experimentation work

and both of them are different from each other in terms of dimensions, as required by the experiment in testing. Since a lot of research has already been done on glass fiber-reinforced composite placed in sea water, hence in this project, specimens are submerged in water, fully saturated salt solution and half-saturated salt solution. The fully saturated salt solution and half-saturated salt solution at room temperature (ambient temperature) 20 °C contains 10.2 times (357 gm. of salt in 1 L of water) and 5.1 times (178.5 gm. of salt in 1 L of the salt) as present in sea water (35 g of salt in 1 L of water roughly equivalent to 35,000 ppm), respectively. These procedures are followed only to evaluate the changes occurred on the composite by a highly concentrated brine solution [12–14].

Kinetics of moisture absorption is studied by placing the epoxy E-glass fiber laminate composite in these solutions. Moisture absorbed by the specimen is regularly checked by taking out the specimen after fixed interval of time. Before taking the weight measurements, water absorbent paper is used to wipe out the surface moisture from the specimen. Electronic weighing machine of 0.0001 gm accuracy is used to measure the weight of specimens.

The gain in moisture content is calculated with the equation below:

$$\text{Absorbed Moisture}(\%) = [(\text{mw}-\text{md}) \times 100]/\text{md} \tag{1}$$

where mw is the wet weight of the specimen while md is the dry weight of the specimen.

2.3 *Charpy Test Characterization on Fabricated Sample*

To investigate the impact behavior on the developed sample, Charpy test has been performed. The standards procedure has been adopted as per the ASTM-23 of Charpy test. The specimen gets fabricated as per the above procedure (Fig. 1) followed by the finishing of the sample as per the required dimensions of the Charpy sample. Figure 4 represents the dimension sketch of the Charpy sample.

3 Results and Discussion

The results of the water absorptions and Charpy test have been analyzed and discussed.

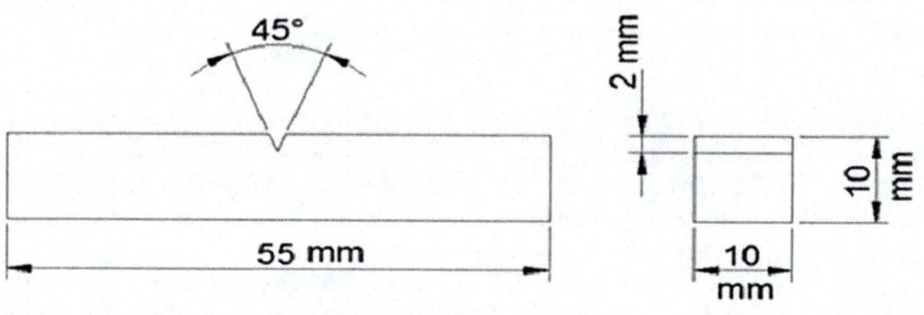

Fig. 4 Dimension sketch of the Charpy test sample

3.1 Water Absorption Test

To specify the result and objective of moisture absorption mechanism in the E-glass fiber laminated composites, respective graphs are plotted between percentages absorbed moisture (Y-axis) and square root of time in hours (X-Axis). The nature and reason behind the selection of such media has been already stated. Tables 3 and 4 represent the recorded data of the water absorption, respectively, for the sample prepared at 43,600 Pa and sample prepared at 24,525 Pa.

Figure 6 represents the graph between the water absorption with respect to the mean time at 43,600 Pa. Similarly, Fig. 7 represents the graphs between the water absorption with respect to the time of a composite sample fabricated at 24,525 Pa. The average data of the similar specimens is represented by each curve. Figure 5 represents the sample before the water absorption test and after the water absorption test for the effective visual.

The rate of moisture absorption varies in different samples in some composites, percentage moisture absorption roughly increases linearly with the increase in immersion time and gradually reaches the equilibrium after some time. The rate of moisture

Table 3 At pressure 43,600 Pa and room temperature

Sr. no.	Square root of time in hours	% Moisture absorption			
		Specimen in air	Specimen in half-saturated salt solution	Specimen in normal water	Specimen in fully saturated salt solution
1	0	0.0	0.0	0.0	0.0
2	9.79	0.0	2.1	2.7	3.1
3	13.85	0.0	2.3	3.9	3.4
4	16.97	0.0	3.0	4.4	4.2
5	19.59	0.0	2.5	4.1	2.7
6	21.9	0.0	2.7	4.6	2.7

Table 4 Pressure at 24,525 Pa and room temperature

S. no.	Square root of time in hours	% Moisture absorption			
		Specimen in air	Specimen in half-saturated salt solution	Specimen in normal water	Specimen in fully saturated salt solution
1	0	0.0	0.0	0.0	0.0
2	9.79	0.0	2.2	2.5	2.8
3	13.85	0.0	2.3	2.7	2.3
4	16.97	0.0	3.1	4.2	3.6
5	19.59	0.0	2.3	3.2	2.6
6	21.9	0.0	2.2	3.6	2.7

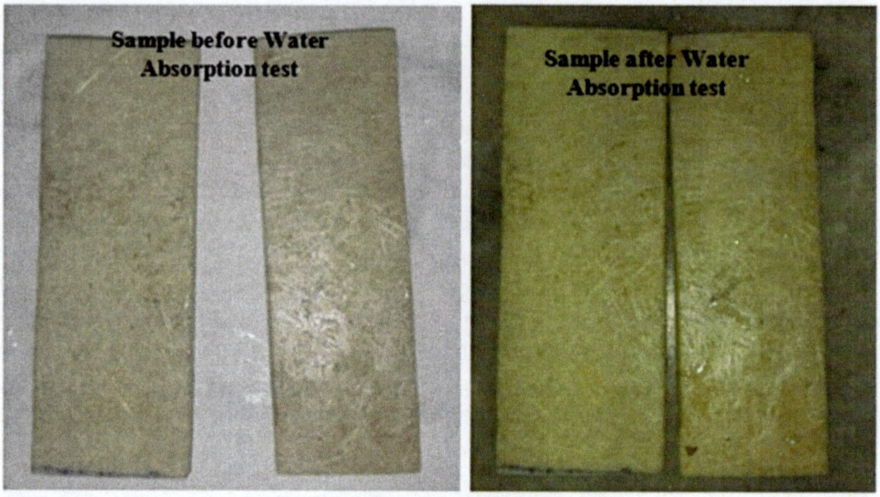

Fig. 5 Sample before water absorption test and after absorption test

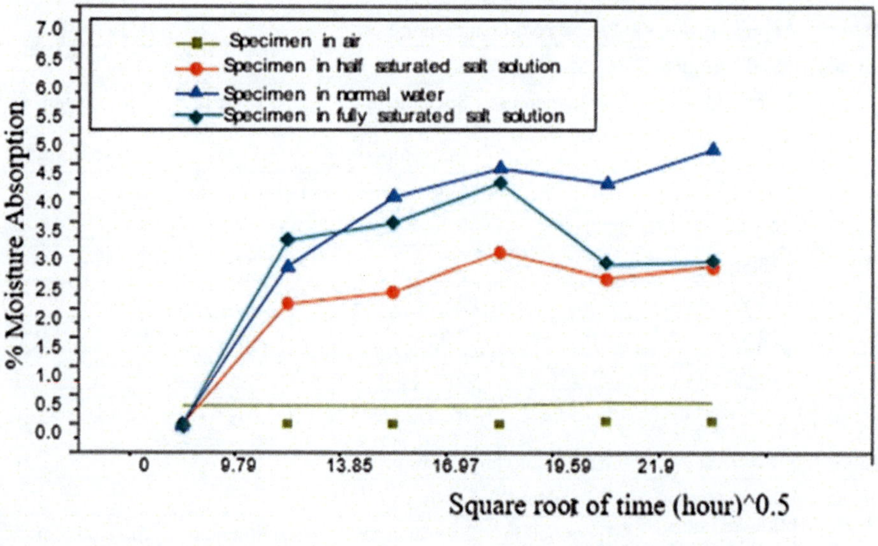

Fig. 6 Moisture absorption curve for sample fabricated at pressure of 43,600 Pa and room temperature

absorption assumes a high value initially followed by a comparatively slower rate moisture absorption as the time passes. This initial increase in moisture content in the composite is commanded by the maximum concentration difference between the composite and medium surrounding the composite. At a particular point, the

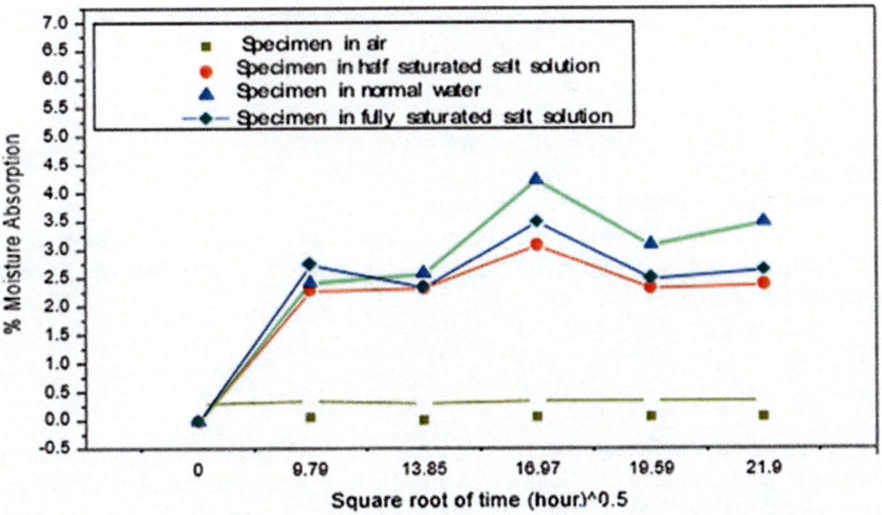

Fig. 7 Moisture absorption curve for sample fabricated at pressure of 24,525 Pa and room temperature

maximum weight is gained by the composite and further negligible moisture absorption or desorption takes place. This kind of behavior marks the presence of the Fickian model based on the Fick's law.

From the theoretical analysis aspects, the time to reach 99% of the maximum moisture content (saturation point) is governed by two important parameters: the maximum moisture content Mm and one-dimensional diffusivity (D). These two parameters are easier to calculate for the moisture diffusion which follows the Fickian model.

Shen and Springer have given the following equation which estimates the moisture content of the material during both adsorption and desorption:

$$M = \mathrm{Mi} + G(M - \mathrm{Mi}) \tag{2}$$

Mi is the initial moisture content of the material, Mm is the maximum moisture content which can be attained under a given environmental conditions, and G is a time-dependent parameter.

$$G = 1 - \frac{8}{\pi^2}\sum_{j=0}^{\infty}\frac{\exp\left[-(2j+1)^2\pi^2\left(\frac{D_x t}{h^2}\right)\right]}{(2j+1)^2}$$

The above Eq. (3) can be approximated as:

$$1 - \exp\left[-7.3\left(\frac{D^{x^t}}{s^2}\right)^{0.75}\right] \tag{3}$$

For a material exposed to two sides under environmental condition, $S = h =$ thickness.

Insulating one side of the same material, $S = 2$ h.

Dx is the diffusivity of the material in the direction normal to the surface.

The time required by a material to attain at least 99.9% of its maximum possible moisture content is:

$$t_m = \frac{0.67\ S^2}{D_x} \tag{4}$$

Graph plotted indicates that the moisture absorbed by E-glass fiber composite is not in accordance with Fickian model. There exists some amount of absorption, desorption, and equilibrium not attained in both the graphs. Such non-Fickian behavior is also reported in past and theoretical calculation of diffusivity and time required for attaining maximum moisture content cannot be estimated by Shen and Springer equation. This has made the prediction of diffusivity comparatively difficult. Cracks on surface, deboning between fiber and viscoelastic behavior in the polymer might result in such variations. Moreover, factors such as curing pressure, binding of salt and water molecules to the resin structure, void content in the resin and the composite might be among some of the other reasons also.

With the help of both graphs, we can comprehend the moisture absorption capabilities of specimen when they are placed in four different aqueous mediums, i.e., normal water, air, half-saturated salt solution, and fully saturated salt solution. From Fig. 4, we can see that the moisture content in air is negligible and we verified it, when the increase in weight of the specimen is found to be very less as in comparison with other solutions.

Due to higher concentration difference, the rate of flow across the boundary for first 4 days, the moisture absorption in specimen immersed in fully saturated salt solution is greater than the moisture absorbed by the specimen in normal water. In case of half-saturated solution, the graph depicts completely different result, and the moisture absorption in the former (43,600 Pa) is less than that of the latter (24,525 Pa) one, even if the concentration difference in the former is comparatively more. This is due to the ingression of normal water into the epoxy polymer is quite more feasible in comparison with saline water that contains dissolved salts. A concentration-driven osmotic pressure causing moisture ingression into the polymeric composite body is less of a problem in saline solutions than in pure water. Here, the ingression of normal water into the composite over-powers the concentration-driven moisture absorption. For fully saturated specimen, moisture absorption dominates.

After first 96 h of immersion, due to the presence of large salt molecules in those salt water solutions, the curve for the specimen in normal water overtakes the other two curves, i.e., the half-saturated and fully saturated salt solutions, in relation to percentage moisture absorption which slows the diffusion process into the matrix of the composite materials, resulting in lower absorption kinetic parameters.

The decrease in the concentration difference is due to the decrease in the slope in the second stage of the weight measurement. However, the curves of the specimens in aqueous medium between the third and fourth stage of the weight measurements show a decreasing trend as each of the three curves attains the maximum moisture content around that particular point of time which is due to inertia of the movement of water in the composite which creates the reverse concentration difference causing reverse osmosis.

Moreover, there is deposition of bulky salt components elements on the surface of composite body intervening with the diffusion processes between the last and second last point of weight measurement, after that there is a gradual increase in the percentage moisture absorption observed following a simple absorption trend. Moisture absorption in these specimens can still attain a maximum value of moisture absorption as the time passes in months or years.

In Fig. 7, curve corresponds to the specimens fabricated at the pressure of the 24,525 Pa and shows trends similar to Fig. 6 curves. After the first 4 days of immersion for specimen in fully saturated salt solution, it is seen that desorption occurs. It could be due to the bonding between the water and the resin, matrix cracking, plasticization behavior of matrix, etc., in fact it is difficult to explain the causes behind such kind of behavior.

This can cause stronger adhesion between the fiber and the matrix which overpowers theoretically estimated percentage void content, and second reason underpinning this nature of curves can be the generation of internal cracking in the former situation due to which such behavior is being noticed. During machining, there were great chances of development of internal cracking in the former situation due to which internal cracking can be noticed.

The half-saturated solution shows more moisture absorption percentage in samples fabricated at 24,525 Pa than samples fabricated at the pressure of 43,600 Pa up to the third stage of weight measurement, and the percentage moisture absorption in the sample fabricated at 43,600 Pa becomes more due to initial swelling of matrix, dominated by the percentage void content, edge effect, and more moisture absorption taking place in the sample at 24,525 Pa while after the time instant of third stage of weight measurement, the mechanical adhesion between fibers and matrix dominates. Therefore, the swelling of the matrix in the case of half-saturated salt solution is much slower than the fully saturated salt solution. This may be due to that there was a distinct amount of swelling in the matrix of both the samples. It may happen because of more swelling in matrix of specimens maintained at 43,600 Pa than the specimens fabricated at 24,525 Pa. The gain in the weight at different time instants depicts the percentage moisture gain in the E-glass fiber composite. There is inapproachable equilibrium till the entire test and non-Fickian model was followed in the E-glass fiber epoxy composite during moisture absorption. Absorption of water continually rises initially in all the samples. And some inequality has been occurred at the interface such as breakdown of the chemical bonding between the glass fiber and the pattern at the interface. Therefore, there is a need to obtain more estimated data

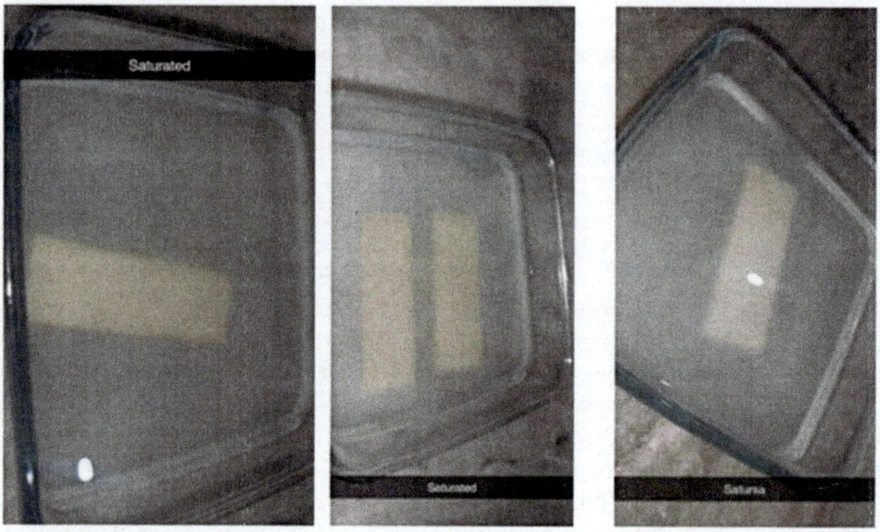

Fig. 8 **i** 99% saturated absorption, **ii** after an ample time. (saturated solution (43,600 pa), and **iii** after an ample time. (saturated solution (24,525 pa)

that can help us in the interpretation of more effects. It was also demonstrable that the flow of moisture absorption in the E-glass fiber epoxy composite can be significantly affected even by minor difference in the recovering phenomena (pressure in this case). Figure 8 represents the visual view of the test procedure.

3.2 Charpy Test Results

Table 5 shows the results of the all sample prepared at varying pressure condition of the hydraulics press. The effect of the varying pressure while fabricating sample was recorded. The results show the clear difference between the samples prepared at different pressure. The sample made up at high pressure shown the higher value of the rupture energy whereas the sample prepared at lower pressure shows the less amount of rupture taken to fracture. The sample prepared at 43,600 Pa shows 8.80% of changes in the rupture energy, similarly the sample prepared at 24,525 Pa shows 4.86% changes.

Table 5 Charpy test results for both the samples prepared at 43,600 Pa and 24,525 Pa

	Specimen fabrication pressure	Specimen condition	Result (Joules)	% Change
Charpy test	43,600 Pa	Specimen in air	58.16	
		Specimen in half-saturated water for square root of 16.97 h	53.04	8.80
		Specimen in full saturated water for square root of 16.97 h	50.07	13.91
	24,525 Pa	Specimen in air	45.25	
		Specimen in half-saturated water for square root of 16.97 h	43.05	4.86
		Specimen in full saturated water for square root of 16.97 h	42.6	5.86

4 Conclusion

Following conclusion has been drawn from the experimentation.

1. The percentage absorption of moisture when it is in air for any time duration is always 0 at 43,600 Pa and 24,525 Pa.
2. If the square root time ranges from 0 to 9.79 percentage absorption at 43,600 Pa start increasing in half-saturated solution, normal water and in fully saturated solution but at 24,525 Pa after a specific time period there is variation in absorption it increases some time and sometimes decreases specifically in half-saturated solution and normal water.
3. The percentage of moisture absorption in samples fabricated at 43,600 Pa and immersed in normal water and fully saturated salt solution is more than the samples fabricated at 24,525 Pa for the entire phase of weight measurements. Now, when more time passes there is a small increase in absorption by 9.52% in half-saturated solution, but there is a large increase in absorption at this time in normal water it increased by 44.44% and a small increase in fully saturated solution, i.e., 9.68%.
4. In fully saturated solution after a time period percentage absorption becomes constant at both the pressures. The percentage void content in the latter case is approximately twice the percentage void content in the former case (43,600 Pa). It indicates that the sample fabricated at the pressure of 24,525 Pa (in the latter case) should have more moisture content.
5. Charpy results reveal that the pressure plays a critical role while fabricating the sample, as the pressure increases the sample particles get closed to each other and make strong bonding in between.

References

1. Adekomaya O, Adama K (2018) Investigating water absorption and thickness swelling tendencies of polymeric composite materials for external wall. Niger J Technol 37(1):167–172
2. Ottolini LM, Almeida JRMD (2018) Evaluation of moisture absorption effects on mechanical behavior of hybrid glass/carbon fibers epoxy material composites. In: Chemical and material engineering department. Pontifical Catholic University of Rio De Janeiro, Brazil, pp 2–7
3. Adekomaya O, Ojo K (2017) Glass-fibre reinforced composites: the effect of fibre loading and orientation on tensile and impact strength. Niger J Technol 36(3):781–787
4. Daramola OO, Akintayo OS, Adewole TA, Talabi HK (2017) Mechanical properties and water absorption behavior of banana fibre composites. In: Department of metallurgical and materials engineering. Federal University of Technology, Nigeria, pp 187–189
5. Muñoz E, García-Manrique JA (2015) Water absorption behavior and its effect on the mechanical properties of flax fibre reinforced bio epoxy composites. In: Department of engineering mechanics and materials. Polytechnic University of Valencia, Spain, pp 45–47s
6. Alamri H, Low IM (2012) Mechanical properties and water absorption behavior of recycled cellulose fiber reinforced epoxy composites. Polym Test 31(5):620–628
7. Holbery J, Houston D (2006) Natural-fiber-reinforced polymer composites in automotive applications. JOM 58(11):80–86
8. Espert FV, Karlsson S (2004) Comparison of water absorption in natural cellulosic fibers from wood and one-year crops in polypropylene composites and its influence on their mechanical properties. Compos Part A Appl Sci Manuf 35(11):1267–1276
9. DeWilde PW, Frolkovic P (1994) The modelling of moisture absorption in epoxies: effects at the boundaries. Composites 25(2):119–127
10. Shen C-H, Springer GS (1976) Moisture absorption and desorption of composite materials. J Compos Mater 10:2–20
11. Dantas CM, Raimundo (2018) Influence of the quantity of water absorption in the mechanical properties of jute fiber and glass fiber composites
12. Dantas CM, Freire Júnior RCS, Santos JKDD, Silva CCD, Cöuras Ford ETL (2018) Influence of the quantity of water absorption in the mechanical properties of jute fiber and glass fiber composites
13. Somani N, Sharma N, Sharma A, Gautam YK, Khatri P, Solomon JAA (2018) Fabrication of Cu-SiC composite using powder metallurgy technique. Mater Today Proc 5:28136–28141
14. Somani N, Gautam YK, Sharma SK, Kumar M (2018) Stastical analysis of dry sliding wear and friction behavior of Cu/SiC sintered composite. In: AIP conference proceedings, p 020018

Four Wheeler's Health Monitoring by Sound Level Measurement: A Case Study

Manpreet Singh, Simran Singh, Rajeev Kumar, Sumit Shoor, Piyush Gulati, Jaiinder Preet Singh, and Harpreet Singh

Abstract Sound level emitted by vehicle components is having great importance in monitoring their health condition. Higher sound level from a vehicle than the specified not only signifies the sign of fault in the main component of automobile but also dangerous to the pedestrians. Over the time, moving parts of vehicle deteriorates and causes the higher level of sound is merely a guess. In this case study, attempt has been made to check the deterioration level of four different Honda city (diesel variant) cars on the basis of their emitted sound levels measured from various components and conditions. The cars were tested for mainly two conditions idle and moving and sound levels were recorded inside and outside the cabin. The cars have been classified on the basis of odometer reading and tested just after their regular service to remove the conditions of service lag. The measured levels of sound clearly state that the car started emitting the high level of sound over the time and can be said deteriorates with its use. The measured sound levels were also proven to be capable of detecting faults in some of the significant components of the vehicle.

Keywords Health monitoring · Vehicle health · Sound level · Sound level monitoring

M. Singh · S. Singh · R. Kumar · S. Shoor (✉) · P. Gulati · J. P. Singh
School of Mechanical Engineering, Lovely Professional University, Phagwara, Punjab 144411, India
e-mail: sumit.14602@lpu.co.in

M. Singh
e-mail: manpreet.20360@lpu.co.in

P. Gulati · J. P. Singh · H. Singh
Department of Planning and Quality, Nettbuss AS, Oslo, Norway

A. K. Dubey et al. (eds.), *Recent Trends in Industrial and Production Engineering*, Lecture Notes in Mechanical Engineering,
https://doi.org/10.1007/978-981-16-3135-1_10

1 Introduction

Sound pollution is term we often come across which is the propagation of disruptive noise in the environment causing a serious impact to the human. The main sources of this disruptive noise close to the human life can be estimated as the traffic noise or the vehicle noise [1–3]. The main concern of noise is not only lying with the manufacturer end to manufacture a vehicle generating less noise but also to the deterioration of the vehicle components over the time. As the vehicle starts aging with the time generally moving components of the vehicle wear out and causes excessive noise [4–7]. This excessive noise generally borne by excessive vibration levels caused due to faulty components [8–12]. There are many techniques available to detect these faults by monitoring the vibration signal and on the basis of that solutions can be derived [13–15]. The five main sources of sound in an automobile are engine, transmission system, cooling system, exhaust system and tires [16]. When engine is turned on and in idle condition then engine sound, partial exhaust sound and partial transmission sound are the mainly contributes to the total sound. Tires and exhaust system are contributing more sound when automobile is in running condition. Each part is having its own significance and reasons for generating sound. By measuring acoustic signal many techniques are available which can detect the faults if it relates with some frequency [17]. The manufacturers are making better designs and using materials having high damping characteristic to absorb sound level to the maximum [18–20]. Even if the sound level of one-part increases then it affects the overall sound level of the automobile.

This study aims to estimate the health condition of four different variants of Honda city (Diesel) with different odometer readings. The sound levels of vehicles were recorded by using sound level meter for the idle and running conditions. All the recorded sound levels were compared for various conditions such as inside cabin and pass-by noise. The condition of the vehicles was explained with the help of recorded sound levels and recommendations were made to the owner.

2 Description of Selected Vehicles

The four selected cars of Honda city (Diesel variant) were categorized as V1, V2, V3 and V4 on the basis of their odometer readings and availability. The car V1 manufactured in year 2018 and driven for 452 kms. The service gap describes how many kms the vehicle was driven after its service. Generally, cars were tested right after their service and the odometer reading was noted right after the sound level recorded. Similarly, other cars with the same data are presented in Table 1 and categorized as V2, V3 and V4 on the basis of odometer reading. V1 has only driven for 452 kms and can be considered as benchmark for comparing the other vehicles.

Table 1 Description of tested cars

Honda city (Diesel variant)	Odometer reading in kms	Last service in kms	Service gap in Kms	Manufacturing year
V1	452	NA	NA	2018
V2	44,821	44,819	2	2018
V3	71,827	71,823	4	2016
V4	124,035	124,030	5	2015

3 Test Results with Discussion

The tests were conducted in three different sets, idle condition, cabin noise while vehicle moving and pass-by noise. In first set of experiment, the vehicles were set to idle condition and various sound levels were recorded. Sound level meter having 35–130 dB range, 0.1 dB resolution and frequency range of 31.5–8000 Hz was used for recording the sound levels. Firstly, engine noise was recorded by keeping 1500 rpm of the engine and without turning on cooling system (Air Conditioner AC) and sound level was recorded after placing the sound level meter at a distance one meter from the engine. The V1 has recorded the sound level of 69.8 dB which is minimum as compared to other vehicles. After this AC unit was turned on and engines speed were kept at 1500 rpm and sound level was recorded in the similar manner. The sound level for this condition for V1 was recorded as 70.3 dB and this sound level is combined sound level of engine and AC unit at 1500 rpm. The sound level of AC unit at 1500 rpm can be calculated as 60.7 dB after deducting the 69.8 dB from 70.3 dB. Similarly, other vehicles were tested and their recorded sound levels are shown in Table 2. In all the conditions V1 has been recorded with the minimum sound levels because the vehicle is new and its condition has not deteriorated due to aging. Table 3 represents the percentage of more sound level generated by other vehicles at the same condition in comparison with the minimum level of V1. The noise levels of vehicles in almost all the conditions are showing the upward trend due to aging factor and it is supposed that with the usage main components get wear out and started producing more vibrations and sound levels. The cooling and exhaust system of V2 has recorded

Table 2 Recorded sound levels for the idle conditions without loading

Conditions @1500 rpm	V1	V2	V3	V4
Engine noise with AC unit turning off	69.8	77.6	78.2	80.2
Engine noise with AC unit turning on	70.3	78.9	78.9	80.7
Cooling system noise	60.7	73	70.6	71
Exhaust sound	66.4	81.4	80.2	80.5
Cabin noise with AC unit turning off	52.1	61.4	62.7	63.5
Cabin noise with AC unit turning on	56.8	63.5	64.2	66.1
Cooling system noise in cabin	55.0	59.3	58.8	62.6

Table 3 Percentage of more sound level generate by other cars in comparison with the minimum level for the idle conditions

Conditions @1500 rpm	V1	V2	V3	V4
Engine noise with AC unit turning off	0	502.5	591.8	996.5
Engine noise with AC unit turning on	0	624.4	624.4	996.5
Cooling system noise	0	**1598.2**	877.2	971.5
Exhaust sound	0	**3062.2**	2298.8	2470.4
Cabin noise with AC unit turning off	0	751.1	1048.1	1280.4
Cabin noise with AC unit turning on	0	367.7	449.5	751.1
Cooling system noise in cabin	0	**169.1**	139.9	475.4

for more sound level than V3 and V4. During inspection, the defect was identified in the radiator fan belt and muffler of V2. The recommendations were made to the owner of the V2 for these defects. The cabin noise levels were also recorded for the similar conditions by turning off and on the AC unit.

In the similar manner, the cooling system sound was calculated by subtracting the sound level without turning on the AC unit from the sound level with turning on the AC unit. The cooling system of V2 is having more sound level than V3 and V4 due to faulty belt.

In the second set of experiments, cars were made to run in the first and second gear at engine speed of 1500 rpm and sound levels were recorded inside cabin by turning on and off the AC unit. The recorded sound levels are presented in Table 4 along with percentage more sound level from the reference sound level of V1 and presented in Table 5. In all the conditions the sound levels are showing the upward trend with the aging of the vehicles. Despite having problem in the fan belt of V2, the cooling system sound for V2 is not recorded for higher values than V3 and V4 in the running conditions. The reason for the same might be due to running condition radiator starts taking air directly and the radiator fan might get turned off. So, the other parts of the cooling system of V2 can be said to be in the right order.

In the third set of experiments, cars were made to run in the first and second gear at 1500 rpm and pass-by sound levels were recorded at various conditions. The pass-by sound levels were recorded by keeping the sound level meter 3 m apart.

Table 4 Recorded sound levels inside cabin for the running cars with various conditions

Cabin noise for moving cars @ 1500 rpm	V1	V2	V3	V4
Cabin noise @ 1st gear with AC unit turning off	0	1521.8	2087.8	2654.2
Cabin noise @ 1st gear with AC unit turning on	0	1188.2	1678.2	2190.9
Cooling system noise in 1st gear	0	771.0	1130.3	1598.2
Cabin noise @ 2nd gear with AC unit turning off	0	751.1	877.2	1598.2
Cabin noise @ 2nd gear with AC unit turning on	0	560.7	658.6	1188.2
Cooling system noise in 2nd gear	0	134.4	169.2	280.2

Table 5 Percentage of more sound level generate by other cars in comparison with the minimum level for the running car recorded inside cabin

Cabin noise for moving vehicles @ 1500 rpm	V1	V2	V3	V4
Cabin noise @ 1st gear with AC unit turning off	51.2	63.3	64.6	65.6
Cabin Noise @ 1st gear with AC unit turning on	53.7	64.8	66.2	67.3
Cooling system noise in 1st gear	50.1	59.5	61.0	62.4
Cabin noise @ 2nd gear with AC unit turning off	55.4	64.7	65.3	67.7
Cabin noise @ 2nd gear with AC unit turning on	57	65.2	65.8	68.1
Cooling system noise in 2nd gear	51.9	55.6	56.2	57.7

The recorded pass-by sound levels at various conditions are presented in Table 6 along with percentage more sound level from the reference sound level of V1 at the same condition and presented in Table 7. The cooling system sound effect outside the vehicle is calculated in the similar manner as done in the previous set of experiments. The V2 is showing excessive sound level in all the cases except cooling system sound than V3 and V4. During inspection, it has been identified that the tires of V2 were more worn out than the tires of V3 and V4. The worn-out tires of V2 were causing excessive vibrations and noise which results in higher sound level recorded for V2 in pass-by conditions. The worn-out tires of V2 were not causing any significant effect

Table 6 Recorded pass-by sound levels for the running cars for various conditions

Pass-by noise for moving cars @ 1500 rpm	V1	V2	V3	V4
Pass-by noise @ 1st gear with AC unit turning off	65.2	74.4	70.8	71.4
Pass-by noise @ 1st gear with AC unit turning on	66.6	75.7	73.6	74.2
Cooling system noise in 1st gear	61.0	69.8	70.3	70.9
Pass-by noise @ 2nd gear with AC unit turning off	64.2	74.6	68.6	69.5
Pass-by noise @ 2nd gear with AC unit turning on	65.6	75.5	71.5	72.2
Cooling system noise in 2nd gear	60.0	68.2	68.3	68.8

Table 7 Percentage of more pass-by sound level generate by other cars in comparison with the minimum level for the running cars

Pass-by noise for moving cars @ 1500 rpm	V1	V2	V3	V4
Pass-by noise @ 1st gear with AC unit turning off	0	**731.8**	263.1	316.9
Pass-by noise @ 1st gear with AC unit turning on	0	**712.8**	401.2	357.1
Cooling system noise in 1st gear	0	658.6	751.1	877.2
Pass-by noise @ 2nd gear with AC unit turning off	0	**996.5**	175.4	238.8
Pass-by noise @ 2nd gear with AC unit turning on	0	**877.2**	289	357.1
Cooling system noise in 2nd gear	0	560.7	576.1	658.6

inside the cabin as observed in the previous set of experiment. The faulty radiator belt of V2 is not showing any signs while in the running conditions.

The sound levels for all the conditions of each vehicle are shown in Figs. 1, 2, 3 and 4 respectively. In all the cases the sound level with turning on the AC unit was coming out to be more due to additional sound addition due to cooling system. The pass-by sound level for engine and cooling system in all the cases was coming out to be more in comparison with the cabin noise. This is because all the vehicles have designed with proper sealing inside the cabin to restrict the sound entering into the cabin. The sound of cooling system was coming out to be more in the 1st gear in comparison with 2nd gear except the cabin noise of V1. This signifies that the cooling system is taking more load in 1st gear in comparison with second gear. The other gear conditions were not tested due to the space restrictions and specified speed limit.

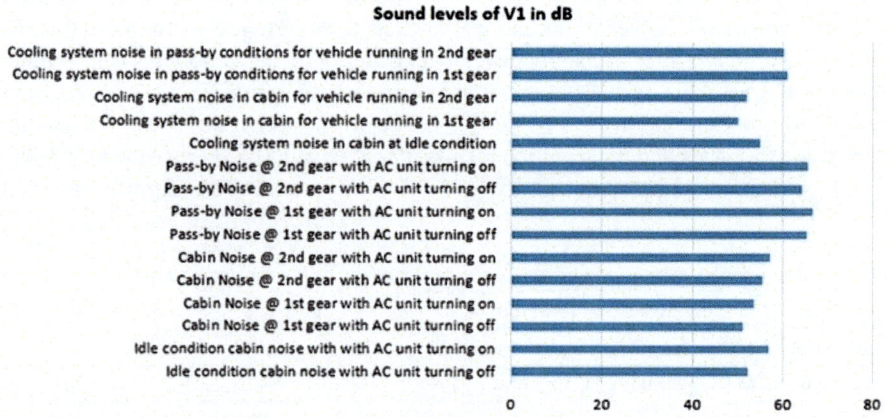

Fig. 1 The recorded sound levels of vehicle 1 for all the conditions

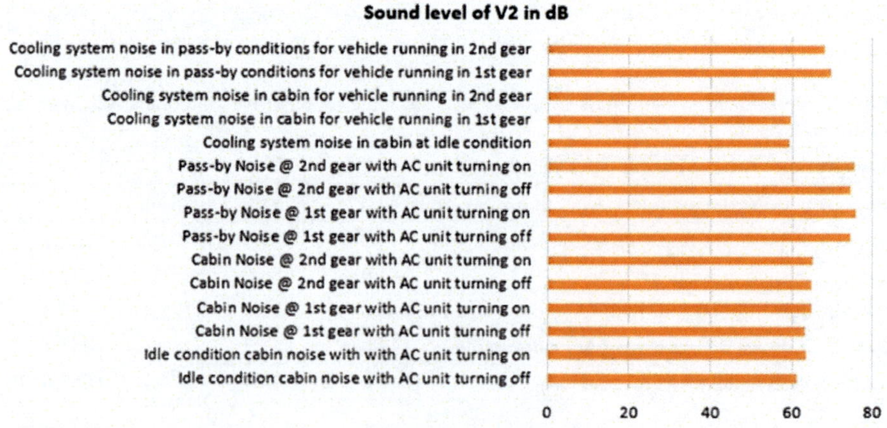

Fig. 2 The recorded sound levels of vehicle 2 for all the conditions

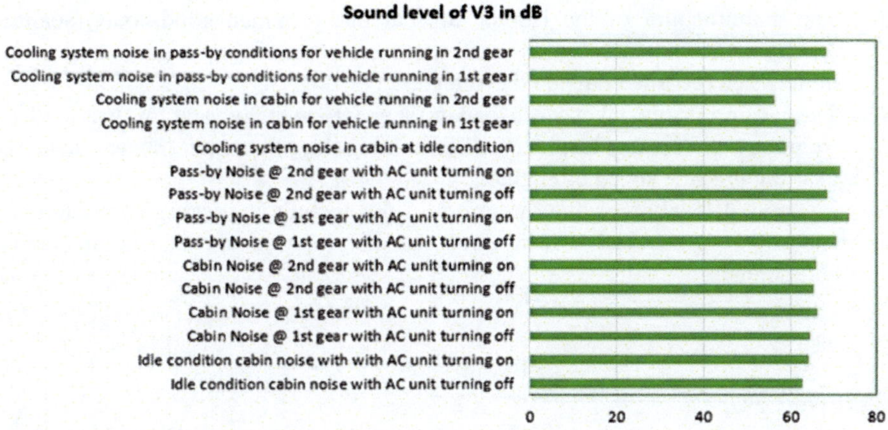

Fig. 3 The recorded sound levels of vehicle 3 for all the conditions

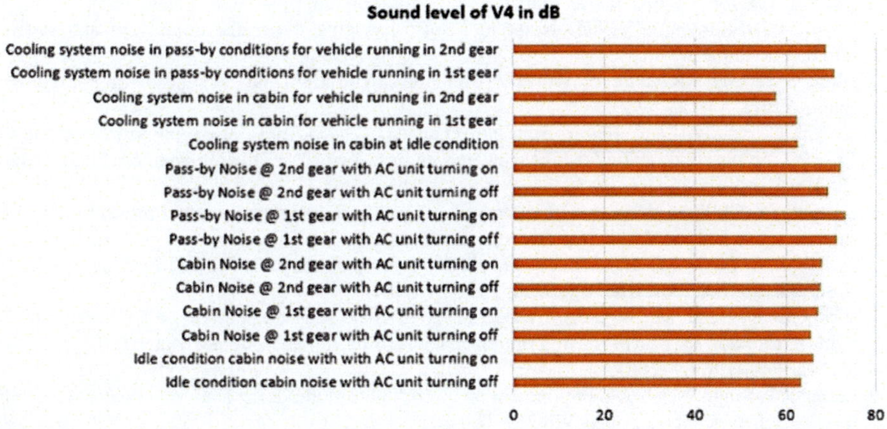

Fig. 4 The recorded sound levels of vehicle 4 for all the conditions

4 Conclusion

The case study done on the Honda city cars has helped in drawing the following conclusion.

1. With the use of vehicles, the moving components deteriorates significantly over time and resulting in emitting higher sound levels. In all the cases the sound level has found to be increased for the components of vehicles with its usage.
2. The sound level of vehicles at various levels has helped in identification of faults in significant components such as radiator fan belt, muffler and worn-out tires.

3. The manufacturer of the Honda city car has provided good sound sealing arrangement for the cabin as in all the cases sound levels were significantly more when recorded outside the vehicles.
4. The cooling system takes more load in 1st gear than in 2nd gear this was evident from the higher sound level recorded for the cooling system in 1st gear than 2nd gear under same speed of engine.
5. The sound level recording system does not include any complex equipment and still it is capable of identifying the deterioration of vehicles and faulty components.

References

1. Karanath NV, Raju S (2005) Investigation of relation between stationary and pass by noise for new in use vehicle. SAE paper no. 2005-26-051. ARAI Pune, pp 623–629
2. Nashif AD, Jones DIG, Henderson JP (1985) Vibration damping. Wiley, New York
3. Vydra EJ, Shorgen JP (1993) Noise and noise reducing materials. Society of automotive engineers. Paper no 931267
4. Raju S (2004) ARAI Pune, Workshop on noise, vibration and harshness for automotive engineering. pp 123–139
5. Nayak N, Reddy PV, Aghav Y, Sohi NS, Dani AD (2005) Study of engine vibration due to piston slop on single cylinder high powered engine. Kirloskar Oil Engines Ltd. Pune. India, SAE paper 2005-26-046, pp 581–588
6. Gosavi SS (2005) Automotive buzz, squeak and rattle (BSR) detection and prevention TATA technologies Ltd", ARAI Pune, pp 661–667
7. Lilley KM, Fasse MJ and Weber PE (2001) A comparison of NVH treatments for vehicle floorplan applications. SAE paper no. 2001-01-1464
8. Cambow R, Singh M, Bagha AK, Singh H (2018) To compare the effect of different level of self- lubrication for bearings using statistical analysis of vibration signal. Mater Today Proceed 5(14), Part 2:28364–28373
9. Kung SW, Singh R (1999) Development of approximate methods for analysis of patch damping and design concepts. J Sound Vib 219:785–812
10. Singh G, Kumar R, Singh M, Singh J (2017) Detection of crack initiation in ball bearing using FFT analysis. Int J Mech Eng Technol 8(7):1376–1382
11. Singh M, Kumar R, Jena DP (2009) Detection of missing ball in bearing using decomposition of acoustic signal. Asian J Chem 21(10):S143-147
12. Kumar R, Jena DP, Singh M (20–22 Sep 2010) Identification of inner race defect in radial ball bearing using acoustic emission and wavelet analysis. ISMA2010 international conference on noise and vibration engineering, Leuven, Belgium, pp 2883–2891
13. Rao MD (2001) Recent applications of viscoelastic damping for noise control in automobiles and commercial airplanes. Keynote speech, Michigan Technological University, Houghton, Michigan & 9931 USA
14. Dere AA, Singh M, Thakan A, Singh H (2019) Structural optimization of Go-kart chassis by geometrical modifications based on modal analysis. ARPN J Eng Appl Sci 14(18):3234–3240
15. Singh M, Shoor S, Singh H (2018) Shannon entropy a better indices for local defect detection and to study the effect of variable loading condition for taper roller bearing. Int J Mech Eng Technol (IJMET) 9(7):198–208
16. Mahale PS, Kalsule DJ, Muthukumar A, Raju S (2005) Vehicle interior noise source identification and analysis for benchmarking, ARAI Pune. SAE paper 2005-26-048, pp 592–603

17. Gabiniemic J, Gatt J, Cerrato G, Jay. (Tecumesh products research laboratory) Automatic detection of BSR events. (Magna Automotive Testing)
18. Soovere J, Drake ML, and Miller VR (1984) Vibration damping workshop proceedings. AFWAL-TR-84-3064 publications by Air force wright aeronautical laboratories, Wright-patterson air force base, Ohio, VV-1-VV-10, a design guide for damping of aerospace structures
19. Kerwin EM (1959) Damping of flexural waves by a constrained viscoelastic layer. J Acoust Soc Am 31:952–962
20. Hussaini A (2001) Designing an interior waterborne coating for use in automotive paint shops to replace sound deadening pads. SAE paper no. 2001-01-1391

Green Manufacturing a Modern Era for Indian Manufacturing Industries: A Review

Jasvinder Singh, Chandan deep Singh, and Dharmpal Deepak

Abstract This paper gives the survey on effectiveness or implementation of green manufacturing in Indian manufacturing industries. This paper focus on the various factors considered for implementation the green manufacturing in industries and analyzes the importance of all these factors. This paper also highlights the use of green manufacturing to form sustainable product and to reuse the product, shorter life cycle. We also tried to discuss the sustainable manufacturing, lean manufacturing and zero waste manufacturing. The main objective of the green manufacturing is to save environment and to reduce the cost of the product.

Keywords Green manufacturing · Lean manufacturing · Sustainable manufacturing

1 Introduction

In present scenario, pollution is increasing due to increasing the number of manufacturing industries. To over the same, green manufacturing plays very important role. It is also known as environmental conscious manufacturing. It also provides the complete information about the environment influence and efficiencies of the resources. In GM, it is targeted that the impact of pollution should be minimum in the environment and utilization of resource should be maximum. Through this attempt, the industries will get profit economically and socially. GM is considered as sustainable strategy in the history of modern manufacturing. Latest GM industries consider green process design, process selection, selection of material, product wrapping, recycling and utilization of resources. GM generally considers the product's total life cycle.

J. Singh (✉) · C. Singh · D. Deepak
Lovely Professional University, Phagwara, India

A. K. Dubey et al. (eds.), *Recent Trends in Industrial and Production Engineering*, Lecture Notes in Mechanical Engineering,
https://doi.org/10.1007/978-981-16-3135-1_11

2 Literature Review

In this section of the paper, attempt has been made to focus on the identification of the important parameters responsible for successful implementation of green manufacturing in Indian manufacturing industries. In this review, approximately 40 papers have been reviewed from four most popular journals from past 15 years.

Ciocci and Percht [1]: The author investigated the electronic industries for implementing the green manufacturing techniques. He also investigated the reaction to environmental regulations for the use of lead-free solders and halogen free flame retardants. In his research, he found same industry successfully implemented the process for getting more profit and to follow the environmental regulations. He also did comparison of environmental regulations with other countries.

Jwahair [2]: The author provided the framework for a holistic evaluation of a product's sustainability value or index in terms of three components of sustainability, i.e., economy, environment and community, over its entire life cycle. This technique is useful in comparing different of the same family's competitive goods. This method uses a visual representation of PSI to give a simple and effective overview of the inherent and integrated levels of sustainability of the product.

Ki-Hoon Le [3]: The author explored and investigated through case study, for same the qualitative methods, interviews and documents for data collection from different companies, one is acoustical equipment, and second is the industry. The study reveals that small and medium-sized industries can be greener by making strategic changes as discussed.

Li [4]: The authors studied that planning and implementation of green manufacturing methodology in Chinese enterprises under the department of Energy Conservation and Pollution Emissions Reduction. The Green Manufacturing Policy needs to be applied in large developed countries with a model of quality improvement. Planning and implementation require a good model in system. It proposes a theoretical model of a structure for planning and implementation of green manufacturing strategy within the context of development.

Feng [5]: The author studied greener industries. His purpose was to explore corporate responsibility in terms of "greenness" by an evaluation of different manufacturing industries response toward consumers. All manufacturing companies are required to transition to "Kantian accountability" with by appointing chief officer and establishment of department related to ethics.

Kung [6]: The study explored the relation between management of green manufacturing and its performance. This was achieved by taking into account any operational aspect includes upstream product discovery and distribution, "research and development, manufacturing and packaging, marketing, promotion and education, and recycling"; they have done a survey and collected data from 188 Taiwanese manufacturers.

Sezen and Cankaya [7]: The author studied the effect of green manufacturing and eco-innovation on sustainability performance of industry. They collected the data from 53 automobile, chemistry and electronics companies through questionnaire base survey and use the methodology to test empirical model by using regression analysis. Through the results of this study, it is explored that green manufacturing applications have significant impact on environmental and social performance. In addition to that, eco-process invention also have a significant impact on the corporate sustainability. However, eco-product invention had not found significant effect on any of the three types of performances.

Tagalpallewar et al. [8]: They observe the questionnaire containing 128 items established on the basis of literature survey and conducting interviews with consulting nine different manufacturing industries, specifically green areas of manufacturing. The contemplate found a total of 12 green manufacturing output counts that were established with their 66 items like top management engagement, information management, staff preparation, green product and process design, staff empowerment, production planning, quality, expense, performance standards for the consumer environment, customer responses and growth of companies. This study explored 107 valid replies from the different industries in India; the limitation of this study is the inadequate sampling.

Vamsi [9]: The author explored that "green product development" which has been appeared as a global occurrence. The main impartial of green product development is to reduce the crash of industrial extension. They perceive out hugely disconnected utilization of elements for executing green product development. The attempt has recognized eighty unique elements and eleven pillars to suggest a framework in the field of "green product development" with the help of relative survey.

Sangwan [10]: The author worked on a topic of green development and related frameworks to find the sources, meanings, scope and similarities. He found that that all the models and strategies have been used by interchanging by various researchers but there should be a standardization in the selection of strategies. During the literature review, it was observed that researchers had to standardize in their study terms researchers must be explicitly explicit on the use of different lifecycle engineering approaches; consistency on end-life methods can be used; incorporation of the whole supply chain and alignment with environmental sustainability initiatives with company strategies.

Fore and Mbohwa [11]: The author observed cement manufacturing sector, a continuous process with the objective to implementing green manufacturing in the process or manufacturing activities. The study analyzed cement industry that have an environmental impact within a developing, low-income country with a specific focus on the sector. In a case study approach, a cleaner production approach was used focusing on issues such as particulate pollution and hazards gases.

Dubey and Ali [12]: The authors examined Indian companies' antecedents of green manufacturing practices and their impact on the performance of the extended supply chain. The data were gathered in two phases, and wave analyses were also conducted to verify non-response bias in order to restrict any significant impact from non-responsive bias on statistical analysis. This data was used to conduct experimentation on factor analysis with varimax rotation. Further use was made of the factor analysis data as an origin of regression analysis.

Orji and Wei [13]: The authors deliberated that manufacturing firms are expected to implement green production and increase the complication of the product at a competitive price. Nevertheless, the discovery of the undertaking costs on green manufacturing is a critical concern for engineering managers. For this reason, an early stage planning and control methodology for green manufacturing costs is important for the engineer.

Prasad et al. [14]: Author explored the applications of green practices to foundry industries in India for improvement in production by reducing waste, by incorporating the GM techniques into work efficiency measures. Thus, findings provide clear evidence that lean and green practices in foundry industries are fairly appropriate for implementation. This paper provides information regarding applicability of lean and green practices in a developing world context and provides proofs that lean and green practices are well applicable in the casting industry.

Seth et al. [15]: The study showed Indian cement industry in context to develop and analyze the green manufacturing-based framework on identified critical success factors and performance measures. The authors adopted data collection system survey. It uses factor analysis of the established critical success factors and regression, along with adequate statistical quality control measures. It provides a comprehensive green manufacturing system connecting performance metrics to top management, management of human resources, corporate culture, good practices, project management and supply chain management.

3 Conclusion

Through this review study, most effective or responsible parameters are found out as shown in Table 1. These parameters can be subdivided in to a greater number of parameters which will be helpful to study the effects of each parameter in quantitative and qualitative way.

Table 1 Green manufacturing drivers

Sr. No	Parameters responsible for implementation of Green manufacturing
1	GM implementation issues
2	Role of legislation in promoting GM
3	Organizational style
4	Eco-knowledge
5	Business environment
6	Society influence
7	Financial incentive
8	Innovation

References

1. Ciocci R (2006) Impact of environmental regulations on green electronics manufacture. Microelectron Int: Int J 45–50
2. Jwahair IS (2006) Total life-cycle considerations in product design for sustainability: a framework for comprehensive evaluation. In: 10th international research conference Barcelona-Lloret de Mar, Spain, Sept 2006, pp 11–15
3. Ki-Hoon Le (2009) Why and how to adopt green management into business organizations? The case study of Korean SMEs in manufacturing iIndustry. Manag Decis 47(7):1101–1121
4. Li K-H (2009) Why and how to adopt green management into business organizations?: The case study of Korean SMEs in manufacturing industry. Manag Decis 1101–1121
5. Feng F-D (2010) "Green" company or "green" consumers: a Kantian retrospective. Int J Soc Econ 779–782
6. Kung F-H (2012) Assessing the green value chain to improve environmental performance: evidence from Taiwan's manufacturing industry. Int J Dev Issues 111–127
7. Sezen B, Cankaya SY (2013) Effects of green manufacturing and eco-innovation on sustainability performance. In: 9th international strategic management conference
8. Tagalpallewar AR, Sunnapwar VK, Digalwar AK (2016)Green manufacturing performance measures: an empirical investigation from Indian manufacturing industries. Measuring Bus Excellence 59–74
9. Vamsi NK (2015) Development of a framework for green product development. An Int J 22(3):246–445
10. Sangwan KS (2015) A bibliometric analysis of green manufacturing and similar frameworks. Manage Environ Qual: Int J 566–587
11. Fore S, Mbohawa S (2015) Greening manufacturing practices in a continuous process industry: case study of a cement manufacturing company. J Eng, Des Technol 94–121
12. Dubey R, Ali SS (2015) Exploring antecedents of extended supply chain performance measures: an insight from Indian green manufacturing practices. Benchmarking: Int J 752–772
13. Orji I, Wei S (2016) A detailed calculation model for costing of green manufacturing. Indus Manage Data Syst 65–86
14. Prasad S, Kinduja D, Sharma SK (2016) An empirical study on applicability of lean and green practices in the foundry industry. J Manuf Technol Manage 2–29
15. Seth D, Shrivastava R, Shrivastava S (2016) An empirical investigation of critical success factors and performance measures for green manufacturing in cement industry. J Manuf Technol Manage 1–29

Investigation of Fused Deposition Modeling (FDM) Process Parameters Influencing the Additively Manufactured Part Characteristics: A Review Paper

Mohit Bhayana, Jaswinder Singh, Bineetpal Singh, and Jaspreet Singh

Abstract The 3D printed parts made by using the additive manufacturing (AM) process are presently high in demand due to many advantages. This includes well-known features such as the cost-effective method of producing customized parts made of thermoplastic material and having a shorter lead time for a wider range of applications. However, it has also a few limitations like poor dimensional accuracy and resolution. Thus, this technology is not suitable for parts having interior details. So, it is very much essential that the parts being produced using this technology must be characterized and analyzed accordingly so as to increase their potential use in the present market. For this, a review paper has been made in order to study all the main characteristics affecting the rise in demand of parts. Several statistical and optimization techniques have been examined and reviewed. The optimal parameters corresponding to the specific characteristics causing an effect to a large extent have been observed.

Keywords Additive manufacturing · Process parameters · Fused deposition modeling · Part quality · Part characteristics · Mechanical properties

1 Introduction

Additive manufacturing (AM) technology, popularly known by its other names such as rapid prototyping and 3D printing, is a production process used for producing the parts by adding layers in such a manner as one layer placed over another consecutively [1]. The G-codes is exported to the computer numerically controlled 3D printing machine so as to form the object [2]. AM finds its applications when producing complex parts required for medium- to small-sized batches of highly demanding personalized and customized parts and products to have shorter lead times with increased throughput levels [3, 4]. Also, AM helps in reducing the wastage of raw material being used for producing the solid objects, because there is only single

M. Bhayana (✉) · J. Singh · B. Singh · J. Singh
Chitkara University Institute of Engineering and Technology, Chitkara University, Punjab, India
e-mail: mohit.bhayana@chitkara.edu.in

A. K. Dubey et al. (eds.), *Recent Trends in Industrial and Production Engineering*, Lecture Notes in Mechanical Engineering,
https://doi.org/10.1007/978-981-16-3135-1_12

operation involved in the building of parts [5]. The art of developing a product rapidly is mainly concerned with the following aspects as development speed, development time, product performance, and development cost [6].

1.1 Methods and Materials

Out of the various 3D printing technologies, FDM 3D printing technology has been found as the field of study because of many advantages like as it can produce neat, clean and complex 3D parts within the stipulated time [7]. The various 3D printing technologies, depending upon the state of input material (solid, liquid, and powder) along with the basic principles on which these technologies are based, are shown in (Fig. 1).

1.2 FDM Process

FDM belongs to the family of material extrusion. In the FDM process, it starts with slicing the CAD data into layers and then this data is imported to the FDM machine. Wherein, the part starts to build up in a layerwise manner as one layer placed onto another over build platform [10]. There is a need of packing the internal structure of the object with the help of outlines or number of contours as per the required response [11].

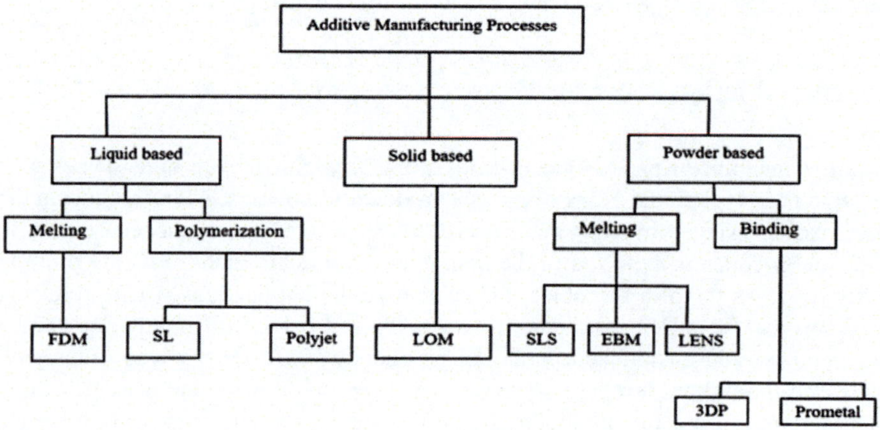

Fig. 1 Classification of additive manufacturing processes [8, 9]

2 FDM Process Parameters Affecting the Characteristics of 3D Printed Parts

The selection of optimized parameters plays a vital role in fulfilling the following qualities of product such as a reduction in material wastage, improvement in quality, high-dimensional accuracy, increased productivity, and reduced production time and cost [12].

2.1 *FDM Process Parameters*

The effect of the selection of process parameters has been observed on the part quality and mechanical properties of a specimen prepared using FDM technique [13, 14]. Table 1 shown below lists the following parameters which are attributed to the characteristics of the specimen made of using FDM 3D printing technology.

2.2 *Mechanical Properties of FDM Printed Parts*

Even though FDM technique has a wide number of applications for which it can be used, but it faces inadaptability to predict its nature of mechanical properties without making the 3D model because of the presence of discontinuity existing in the form of voids in the structure of part [17]. While working on FDM machine for the processing of PEEK material, values of parameters affecting the specimen's properties must not be taking either too high or too low because it may be responsible for affecting the

Table 1 List of FDM process parameters [15, 16]

Parameters	Description
Build orientation (°)	It gives the information about how the part would be printed inside the build platform or it decides how the part, would be positioned inside build platform concerning x-, y-, and z-axis
Layer thickness (mm)	The layer getting deposited overprinting platform after extrusion from nozzle tip forming a thickness or height of layers is known as layer thickness and measures in mm
Air gap	It is generally defined as the gap between the raster tools paths located adjacent to each other on the same layer
Raster angle	It is the angle made by the raster pattern on the bottom layer along with x-axis. It typically varies from 0° to 90°
Raster width	It is defined as the width of the deposited material bead along the path of the extruder tool. Higher the value of raster width causes a stronger interior of the part

surface finish in a poor manner [18]. The mechanical properties also affect the level of crystallinity, which is found to be affected by the printing temperature [19].

Tensile Strength of Printed Specimens Using FDM

Parts to have kept at 0° build orientation over build platform, lying down flat in x–y plane, show the maximum tensile strength [20]. The raster angle also causes to noticeable change in the tensile strength of specimen. The parts with 0° raster angle have been examined to have tensile strength more than the parts where fibers arrange themselves to the applied load direction in parallel manner. "Parameters such as low layer thickness and high raster width result in more bonding area, thereby improving the tensile strength of specimens [21]."

Compressive Strength of Printed Specimens Using FDM

From the past published literature, it has been found that higher the layer thickness results in the higher compressive strength of the specimen [22].

Impact or Fracture Strength of Specimens

The study done in this regard indicated that impact strength was found to be affected certainly with many parameters such as percent infill, infill pattern, number of contours, and air gap. For 100% infill, it was found to be at its maximum value [23]. It was reported that the specimen has a higher impact strength for the infill pattern as crisscross [24].

Flexural properties of Specimens

The various input parameters like air gap, contour width, and number of contours, etc., affect the flexural strength of specimens up to a large extent. However, raster angle and raster width parameters have influenced the most to the flexural strength [25].

2.3 Part Quality Characteristics

The smaller value of part orientation and layer height along with the larger value of raster angle reduces wear [26]. From the viewpoint of literature study for the surface finish, it was found that layer height significantly affects the surface finish of the part [27].

Build Time/Printing Time Quality Characteristics

Filling velocity and layer height found to be the significantly affecting parameters to the build time [28].

Dimensional Accuracy Type Characteristics

The effect of extrusion temperature has been observed on dimensional accuracy of the specimen [29]. If this is too high, it causes material degradation.

3 Results and Discussions

Referring to Table 2 discusses the literature survey conducted up to date for optimum parameters selection in the FDM process corresponding to the output characteristics by utilizing different DOE and optimization techniques.

4 Conclusions

The characteristics of thermoplastic materials like PLA, ABS, PC, etc., based on products and the input parameters affecting the characteristics of these thermoplastic-based parts have been studied till now by various researchers. However, there are very few studies on materials such as nylon and composite. Moreover, the attention has been mainly focused only upon the mechanical properties of the specimen made up of using the thermoplastic material. The properties of a specimen such as tensile, compressive, and flexural or bending strength were studied. The input parameters studied were very few. Therefore, there could be a scope to analyze properties with some more parameters (environment factors) that have not been yet tested. Several optimization techniques have been used to find out the optimal process parameter selection and settings as well for the required characteristics of specimen printed using FDM. Thus, there is a need to identify the optimal selection and settings of input parameters required for multipurpose response optimization technique in order to simultaneously investigate the number of outputs at a time.

Table 2 Literature survey discussing optimum parameters selection corresponding to output characteristics

Author and year	Research objective	Selected FDM input parameters for optimization	Optimization technique/method or tool	Optimum input parameters setting corresponding to output characteristic of part
Torres et al. [30]	Effect of input parameters on FDM specimen in shear with simultaneous optimized mechanical properties	Layer thickness, infill density, and post-processing heat treatment time at 100 °C	Taguchi L9 OA, regression analysis and ANOVA	Highest shear modulus −1265 MPa at layer thickness −0.1 mm, infill −100%, and post-processing heat treatment time −20 min
Zaldivar et al. [31]	To study the influence of processing and orientation print effects on the mechanical and thermal behavior of specimen made of ULTEM-9085 material via FDM 3D printer	Build or print orientation	Digital image correlation (DRC) technique	1. The part standing on the edge of other printed part lying flat in the x–y plane is the strongest specimen
Liu et al. [32]	To study the influence of input parameters on mechanical properties of specimen using ANOVA and gray relational analysis as an optimization technique	Deposition or build orientation, deposition style, thickness of the layer, raster width, and air gap	Taguchi L27 OA method, ANOVA, S/N ratio, and gray relational analysis	Flexural strength, impact strength and tensile strength as 50.34 MPa, 83.51 MPa and 23.07 MPa, respectively, at build orientation, layer thickness, deposition style, raster width and raster interval to be 0°, 0.3 mm, 0, 0.5 mm and −0.1 mm, respectively

(continued)

Table 2 (continued)

Author and year	Research objective	Selected FDM input parameters for optimization	Optimization technique/method or tool	Optimum input parameters setting corresponding to output characteristic of part
Raju et al. [33]	Finding out the application of a hybrid PSO–BFO evolutionary algorithm for optimization of FDM process parameters	Layer thickness, material as a support, part orientation, solid interior	Taguchi L18 orthogonal array (OA)	For all output responses, the input parameters were selected as layer thickness −0.07 mm, support material −sparse, part or build orientation (°) −60, and model internal −high density
Srivastava and Bhaskar [34]	To experimentally investigate the effect of printing parameters of FDM-based 3D printers on average surface roughness	Layer thickness, temperature of the filament at nozzle and print speed	Taguchi L9 OA, ANOVA, response surface methodology (RSM)	Highest roughness was observed at layer thickness −0.3 mm, nozzle temperature −210 °C, and printing speed −60 mm/s
Deng et al. [35]	Optimization of mechanical properties for the poly-ether-ether-ketone-based specimen via FDM	Print temperature, layer thickness, printing velocity, feed rate	Taguchi L9 OA, range analysis	1. Optimum tensile strength as 40 MPa observed at printing speed − 60 mm/s, printing temperature −370° C, filling rate −60%, and layer thickness −0.25 mm.2. For optimum elongation rate as 14.3% at printing speed −20 mm/s, printing temp. −370 °C, layer thickness − 0.25 mm, and filling rate −40%

(continued)

Table 2 (continued)

Author and year	Research objective	Selected FDM input parameters for optimization	Optimization technique/method or tool	Optimum input parameters setting corresponding to output characteristic of part
Fitzharris et al. [36]	To study the effect of improvement in interlayer bonding of material extrusion parts made of polyphenylenesulfide using the Taguchi method	Print temperature, heat treatment of time and temperature	Taguchi L9 OA method, S/N ratio analysis	For percent crystallinity and Young's modulus, parameters were observed to be as heat treatment time −10 min, print temperature −300 °C, and heat treatment temperature −180 °C
Zaman et al. [37]	To study the impact of (FDM) process parameters on strength of built parts using Taguchi's design of experiments	Layer thickness, number of contours, infill pattern, infill percentage	Taguchi L8 OA method, ANOVA, S/N ratio analysis	The combination of parameters such as layer thickness −0.2 mm, number of contours −4, diagonal as the infill pattern, and percentage of the infill as 70% was found to increase the compressive strength to its optimal value

References

1. Zhang J, Jung YG (eds) (2018) Additive manufacturing: materials, processes, quantifications and applications. Butterworth-Heinemann
2. Hashmi S (2014) Comprehensive materials processing. Newnes
3. Ahuja B, Karg M, Schmidt M (Mar 2015) Additive manufacturing in production: challenges and opportunities. In: Laser 3d manufacturing II, vol 9353. International Society for Optics and Photonics, p 935304
4. Whitney DE (1988) Manufacturing by design. Harv Bus Rev 66(4):83–91
5. Vayre B, Vignat F, Villeneuve F (2012) Metallic additive manufacturing: state-of-the-art review and prospects. J Mech Industry 13:89–96
6. Bernard A, Fischer A (2002) New trends in rapid product development. CIRP Ann Manufact Technol 51(2):635–652
7. Srivastava M, Rathee S (2018) Optimization of FDM process parameters by Taguchi method for imparting customized properties to components. Virtual Phys Prototyping 13:203–210
8. Kruth JP (1991) Material incress manufacturing by rapid prototyping techniques. CIRP Ann 40(2):603–614
9. Wong KV, Hernandez A (2012) A review of additive manufacturing. ISRN Mech Eng 2012:1–10
10. Turner BN, Strong R, Gold SA (2014) A review of melt extrusion additive manufacturing processes: I. Process design and modeling. Rapid Prototyping J
11. Ning F, Cong W, Wei J, Wang S, Zhang M (June 2015) Additive manufacturing of CFRP composites using fused deposition modeling: effects of carbon fiber content and length. In: International manufacturing science and engineering conference, vol. 56826. American Society of Mechanical Engineers, p V001T02A067
12. Mohamed OA, Syed HM, Jahar LB (2015) Optimization of fused deposition modeling process parameters: a review of current research and future prospects. Advan Manuf 3(1):42–53
13. Masood SH (1996) Intelligent rapid prototyping with fused deposition modeling. Rapid Prototyping J
14. Groza JR, Shackelford JF (2010) Materials processing handbook. CRC Press, Boca Raton
15. Dey A, Yodo N (2019) A systematic survey of FDM process parameter optimization and their influence on part characteristics. J Manuf Mater Process 3:64
16. Moradi M, Meiabadi S, Kaplan A (2019) 3D printed parts with honeycomb internal pattern by fused deposition modeling; experimental characterization and production optimization. Met Mater Int
17. Coogan TJ, Kazmer DO (2017) Healing simulation for bond strength prediction of FDM. Rapid Prototyping J 23(3):551–561
18. Deng X et al (2018) Mechanical properties optimization of poly-ether-ether-ketone via fused deposition modeling. Materials 11(2):216
19. Sun X, Cao L, Ma H, Gao P, Bai Z, Li C (2017) Experimental analysis of high temperature PEEK materials on 3D printing. In: Proceedings of the 2017 9th international conference on measuring technology and mechatronics automation. Changsha, China
20. Sood AK, Chaturvedi V, Datta S, Mahapatra SS (2011) Optimization of process parameters in fused deposition modeling using weighted principal component analysis. J Adv Manuf Syst 10(2):241–259
21. Rajpurohit SR, Dave HK (2018) Analysis of tensile strength of a fused filament fabricated PLA part using an open-source 3D printer. Int J Adv Manuf Technol 4:1–12
22. Shubham P, Sikidar A, Chand T (2016) The influence of layer thickness on mechanical properties of the 3D printed ABS polymer by fused deposition modeling. In: Key engineering materials, vol 706. Trans Tech Publications Ltd., pp 63–67
23. Alvarez KL, Lagos RF, Aizpun M (2016) Investigating the influence of infill percentage on the mechanical properties of fused deposition modelled ABS parts, Ing e Inv 36(3):110–116

24. Fatimatuzahraa AW, Farahaina B, Yusoff WAY (2011) The effect of employing different raster orientations on the mechanical properties and microstructure of fused deposition modeling parts, business, engineering and industrial applications (ISBEIA), IEEE symposium. pp 22–27
25. Gebisa AW, Lemu HG (2018) Investigating effects of fused-deposition modeling (FDM) processing parameters on flexural properties of ULTEM 9085 using designed experiment. Materials 11(4):500
26. Sood AK, Ohdar RK, Mahapatra SS (2012) Experimental investigation and empirical modeling of FDM process for compressive strength improvement. J Adv Res 3(1):81–90
27. Singh R, Singh S, Singh IP, Fabbrocino F, Fraternali F (2017) Investigation for surface finish improvement of FDM parts by vapor smoothing process. Compos B Eng 111:228–234
28. Dey A, Yodo N (2019) A systematic survey of FDM process parameter optimization and their influence on part characteristics. J Manuf Mater Process 3(3):64
29. Ouballouch A, Ettaqi S, Bouayad A, Sallaou M, Lasri L (2019) Evaluation of dimensional accuracy and mechanical behavior of 3D printed reinforced polyamide parts. Procedia Struct Integrity 19:433–441
30. Torres J, Cotelo J, Karl J et al (2015) Mechanical property optimization of FDM PLA in shear with multiple objectives. JOM 67:1183–1193
31. Zaldivar R, Witkin D, McLouth T, Patel D, Schmitt K, Nokes J (2017) Influence of processing and orientation print effects on the mechanical and thermal behavior of 3D-printed ULTEM 9085 Material. Addit Manuf 13:71–80
32. Liu X, Zhang M, Li S, Si L, Peng J, Hu Y (2017) Mechanical property parametric appraisal of fused deposition modeling parts based on the gray Taguchi method. Int J Adv Manuf Technol 89:2387–2397
33. Raju M, Gupta MK, Bhanot N, Sharma VS (2019) A hybrid PSO–BFO evolutionary algorithm for optimization of fused deposition modeling process parameters. J Intell Manuf 30(7):2743–2758
34. Srivastava A, Bhaskar J (2020) Experimental investigations of printing parameters of fused deposition modeling-based 3D printers for average surface roughness. In: Advances in additive manufacturing and joining. Springer, Singapore, pp 253–265
35. Deng X, Zeng Z, Peng B, Yan S, Ke W (2018) Mechanical properties optimization of poly-ether-ether-ketone via fused deposition modeling. Materials 11:216
36. Fitzharris ER, Watt I, Rosen DW, Shofner ML (2018) Interlayer bonding improvement of material extrusion parts with polyphenylene sulfide using the Taguchi method. Addit Manuf 24:287–297
37. Zaman UKU, Boesch E, Siadat A et al (2019) Impact of fused deposition modeling (FDM) process parameters on strength of built parts using Taguchi's design of experiments. Int J Adv Manuf Technol 101:1215–1226

Investigation of Manual Material Tasks Performed by Workers in North Indian Manufacturing Industries

Jaswinder Singh, Kulwinder Singh, K. Z. Molla, Rakesh Goyal, and Rupesh Gupta

Abstract Lifting load more than the physical lifting capacity results in increase of work-related musculoskeletal disorders among workers. The objective of this study was to investigate the effect of manual material handling (MMH) tasks on musculoskeletal disorders observed by workers in manufacturing industries. A survey on worker's musculoskeletal disorders was carried out using standard Cornell musculoskeletal discomfort questionnaires (CMDQ). The study population consisted of 206 workers from different industries in North India doing material handling work. A total 81.55% ($n = 168$) of workers reported that they had been troubled with musculoskeletal symptoms in one or more of the twelve defined body parts during the last 12 months. Results showed that due to a lack of standards for optimum load for MMH tasks, workers were exposed to musculoskeletal disorders. Lower back, wrist, shoulder, neck, upper back were the most painful problem areas which were observed in the workers.

Keywords Cornell musculoskeletal discomfort questionnaires (CMDQ) · Manual material handling (MMH) tasks · Work-related musculoskeletal discomfort

1 Introduction

Manual material handling (MMH) tasks continue to prevail in manufacturing industries which include the tasks like lifting or lowering, carrying, and holding of a load by the worker. MMH tasks have long been accepted as a major contributor to the occurrence of work-related musculoskeletal disorders among workers [1, 2]. The maximum work-related pain occurs in low back area of human body due to MMH tasks [3]. MMH tasks may also result in musculoskeletal disorders in other body regions. Musculoskeletal disorders (MSDs) are injuries and disorders that affect our musculoskeletal system, i.e., muscles, tendons, ligaments, nerves, disks, etc. Heavy physical work demands place the worker under undue physical stress and increase the

J. Singh (✉) · K. Singh · K. Z. Molla · R. Goyal · R. Gupta
Chitkara University Institute of Engineering and Technology, Chitkara University, Punjab, India
e-mail: Jaswindersingh@chitkara.edu.in

A. K. Dubey et al. (eds.), *Recent Trends in Industrial and Production Engineering*, Lecture Notes in Mechanical Engineering,
https://doi.org/10.1007/978-981-16-3135-1_13

possibility of the inception of work-related musculoskeletal disorders (WRMSDs) [4]. According to the Bureau of Labor Statistics, musculoskeletal injuries are among the most prevalent and costly of all lost time injuries in almost every industry [5]. The workplace risk factors like heavy manual material handling, lifting, heavy physical work, repetitive work, frequency and duration of tasks, age of worker, gender, anthropometries like body weight and person height are the most important causes of work-related musculoskeletal disorders [6]. In recent year's investigations, work-related musculoskeletal disorders have attracted considerable attention because of its importance in assessing ergonomics risk factors involved in manufacturing industries. Many studies have been already done on different types of working conditions. As an example in a study, it was concluded that in steel-making industries, the prevalence of low back pain ranked top, regardless of job titles followed by knee pain [7]. Through correlation analysis, the relationship and interactional features between comfort level, environmental factors (noise, heat, and lighting), workstation dimension, and activities performed by the automotive industries operators while working were determined which results that working environment caused heat/sweat is the most significant discomfort in the job satisfaction [8]. In a study on packaging industry workers, it was concluded that 44.7%, 36.8%, 31.6% of workers had been suffering from knee, back, and neck pain, respectively [9]. It was concluded in a survey at plastic manufacturing industry that upper limb musculoskeletal discomforts were related to work repetitiveness, psychosocial demands, job dissatisfaction, gender and physical unfitness [10]. It was concluded that for 10 and 15 kg load, there is decrease in task time and physical stress associated with lifting. For lifting more than 20 kg, it is recommended that the load should be shared between two workers [11]. The workers in manual wall coating jobs felt discomfort in various body parts during their work [12]. The workers in rubber industry were found to be under symptoms of high severe neck pain [13]. Ergonomics awareness among workers performing MMH activities was done to find out the discomfort faced by workers involve in various manual material activities [14]. Impact of age on lower back performed during MMH tasks was studied. In another study, ergonomics intervention in MMH tasks was done to find out the work-related low back pain [15]. To design and/or analyze manual material handling tasks, there are several observational, direct methods and guidelines available [16]. The guidelines proposed by NIOSH are used mostly and adopted universally to design or analyze MMH tasks [17]. The lifting equation developed by NIOSH is used for the assessment of MMH tasks. This equation was established as supporting assistance for evaluating the physical stress resulting due to MMH tasks [3]. Studies on manual load lifting tasks for the Indian male population have been relatively few. So there is a need to review lifting task with respect to the Indian male population. The objective of this study was to investigate the manual material tasks and its effects on musculoskeletal disorders among workers performed in manufacturing industries, which directly or indirectly affects the worker performance, efficiency, and productivity. This directly put the undesirable expenses on manufacturing cost and having inverse effect on company's growth. So it is important to overcome or decrease this problem.

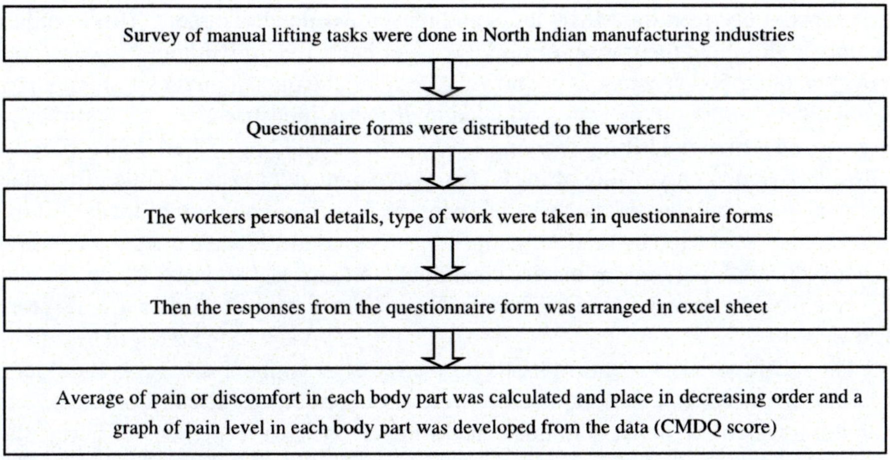

Fig. 1 Flowchart for the procedure adopted for the research

2 Material and Methods

In this research paper, a survey was carried out on manual lifting tasks (MMH) in North Indian manufacturing industries. The survey consists of two parts: first part consists of general detail of workers (like name, age, working experience) and second part consists of finding musculoskeletal discomfort among workers by using Cornell musculoskeletal discomfort questionnaires (CMDQ). The flowchart for the procedure adopted for the research is shown in Fig. 1.

The CMDQ divides the human body (viewed from the back) into twelve anatomical regions in which the question "score the severity of ache, pain, or discomfort in the body part and interference of the pain and effect on the work" was asked for each anatomical area in turn. The discomfort score ranged from 0 to 10 (0 for never 1.5 for 1 to 2 times last week, 3.5 for 3 to 4 times last week, 5 for every day, and 10 for many times a day). This value is multiplied by uncomfortable level (1 for slightly uncomfortable, 2 for moderately uncomfortable, and 3 for very uncomfortable) and interference in workability score (1 for not at all, 2 for slightly interfered, and 3 for substantially interfered). The NIOSH 1991 equation was used to find out the recommended weight limit (RWL) and lifting index (LI) based on the observations for MMH tasks [3].

3 Results and Discussion

A self-reported evaluation using a questionnaire was done for assessing the work-related musculoskeletal disorders among workers due to MMH tasks in manufacturing industries. Most studies on MMH tasks found out that there is a significant relationship between the MMH tasks and musculoskeletal disorders. These studies mainly highlighted the impact of it in the lower back area of workers. However, the effect on other body regions is unknown. A total of 206 questionnaires were filled by the workers, and a survey was conducted in different industries. The tasks surveyed were manual material lifting, carrying, push, pull, and holding of load. Thirty percent of tasks included only lifting of loads. Forty-two percent of tasks include lifting and carrying of loads. Rest of the tasks includes push/pull and holding of loads. All the workers were selected by random sampling. All respondents were male workers and mean working experience as material handling worker was 7.97 years. Mean age was 30 years. No worker was under 18 years of age. The youngest worker was 18 years old.

The self-reported discomfort levels for the lower and upper back, neck, shoulders, elbows, wrist/hands, knees, and elbow were shown in Fig. 2. The discomfort levels for the lower back, wrist, shoulder, neck, upper back were higher. However, the discomfort levels for the elbow, knee, upper arm, thigh, ankle/feet were lower. The result of the study shows that lower back pain was found to be the most common problem in the studied population with a mean severity rate of 11.93 shown in Fig. 2 as 81.55% ($n = 168$) of sample population was suffering from the lower back pain. The severity rate of back was highest in workers who were doing only lifting tasks. Second most common severe pain factor was wrist pain with mean severity rate of 4.28; third most common factor was shoulder pain with mean severity rate of 3.53 followed by neck pain with mean severity rate of 3.28, lower leg with mean severity rate of 2.10, upper back pain with mean value of 1.76, elbow pain with mean value of 1.54, upper arm pain with mean severity rate of 1.27, thigh pain with mean severity

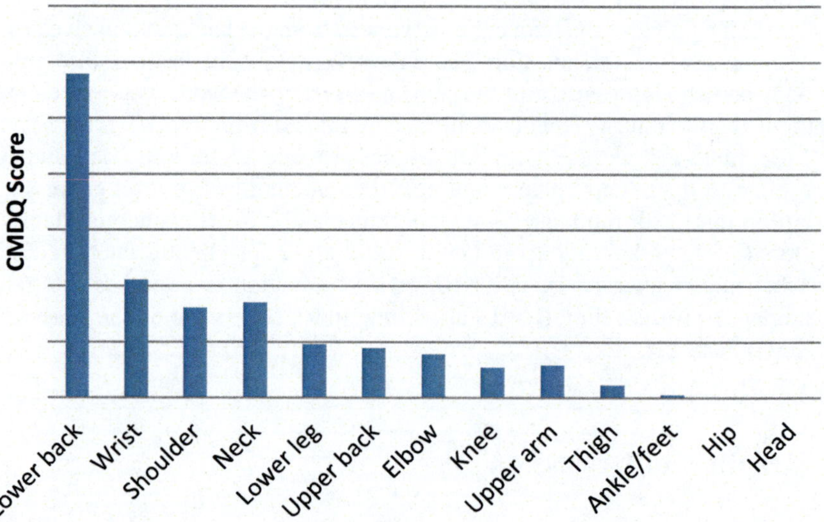

Fig. 2 Mean of CMDQ score in various body parts

rate of 0.70, and ankle/feet pain with mean severity rate of 0.25 were observed in the workers which were exposed to manual material handling.

From the observations in industries, it was found that for some tasks lifting index found by NIOSH 1991 equation was more than 2 and 3. According to NIOSH, MMH tasks with lifting index of more than 1.0 is a potential contributor for work-related discomfort resulting mainly in low back pain. Workers will be at higher possibility of getting work-related discomfort when executing high demanding lifting tasks, where the lifting index is more than 3.0.

4 Conclusions

From the result of this study, it can be concluded the MMH tasks have a significant effect on work-related musculoskeletal disorders. It was found that the symptoms of musculoskeletal disorders are common among most industrial workers, mainly in their lower back. Results show that the evaluation of CDMQ score gives symptoms of work-related musculoskeletal disorders which can be used as an indication of uneasiness in the related body area.

References

1. Chaffin DB (1987) "Biomechanical aspects of workplace design" Handbook of human factors, 1st edn. Wiley, New York
2. Mital A (1983) The psychophysical approach in manual lifting—a verification study. Hum Factors 25:485–491
3. Waters T, Puts-Anderson V, Garg A, Fine L (1993) Revised NIOSH equation for design and evaluation of manual lifting tasks. Ergonomics 36:749–776
4. OSHA, Ergonomics, 2002. Retrieved from http://www.osha.gov/SLTC/ergonomics/index.html
5. Bureau of Labor Statistics, Industry injury and illness Data, 2002. Retrieved from http://www.bls.gov/iif/oshsum.htm
6. Ayoub MM, Dampsey PG, Karwowski W (1997) Manual material handling. In: Handbook of human factors and ergonomics, 2nd ed. Wiley, New York, pp 1085–1123
7. Habibi E, Fereidan M, Mollaaghababai A, Pourabdian S (2007) Prevalence of musculoskeletal disorders and associated lost work days in steel making industry. Iranian J Pub Health 37(1):83–91
8. Ismaila AR (2008) The relation between the discomfort level of automotive industries operators towards their workstation design and work environment. J Achieve Mater Manuf Eng 3(2).
9. Sakineh V, Safari VA, Mohammadi ZI, Jahani HH (2009) Evaluation working posture and musculoskeletal disorders prevalence in pharmacy packaging workers. Euro J Sci Res 29(1):82–88
10. de Cassia Pereira FR, Avila AA, Silvany NAM, Martins CF (2010) Musculoskeletal disorders among workers in plastic manufacturing plants. Rev Bras Epidemio 13(1):11–20
11. Sarbjeet S, Sunand K (2010) The effect of mechanical lifting aid in single task lifting using revised NIOSH lifting equation. Int J Adv Eng Technol 1(2):165–172
12. AbdRahman MN, Rani A, Rebi M, Rohani M (2010) Investigation of the physical risk factor in wall plastering job using WERA method. In: World Congress on engineering 2012, vol 2189, July 4–6, 2012. London, pp 1302–1304. International Association of Engineers.

13. Chow-Li S, Mohd Yusoff BA, Binti Abd RA, Syed HST, IBin K (2012) Prevalence of neck pain and associated factors with personal characteristics, physical workloads and psychosocial among male rubber workers in FELDA settlement Malaysia. Global J Health Sci 4(1)
14. Deros BMd, Daruis DDI, Basir IM (2015) A study on ergonomic awareness among workers performing manual material handling activities. Proc-Soc Behav Sci 195:1666–1673
15. Shojaei I et al (2016) Age related differences in mechanical demands imposed on the lower back by manual material handling tasks. J Biomech 49(6):896–903
16. Wurzelbacher SJ et al (2020) The effectiveness of ergonomic interventions in material handling operations. Appl Ergon 87:103139
17. Rajesh R, Maiti J, Reena M (2018) Decision tree for manual material handling tasks using WEKA. Ergonomic design of products and work systems-21st century perspectives of Asia. Springer, Singapore, pp 13–24

Investigation of Tribological Properties of Stir Cast Hybrid Aluminum Composites

Sunil Kumar Tiwari, Shashank Pal, Abhishek Sharma, Ankit Dasgotra, and Jitendra Kumar Pandey

Abstract The need of enhanced properties of metal matrix composites for aerospace and automotive industries has led to the fabrication of hybrid composites. Use of ceramic particles for fabricating metal matrix composites has been in the eye of researchers. Many researchers have fabricated ceramic particles-reinforced aluminum composites to study their mechanical and tribological properties. This study reflects the effect of reinforcement (boron carbide and silicon carbide) and different loading conditions, on tribological properties of hybrid aluminum composite. Composite samples, $Al - 0wt\%B_4C + 0wt\%SiC$ (non-reinforced), $Al - 2.5wt\%B_4C + 2.5wt\%SiC$, and $Al\text{-}5wt\%B_4C + 5wt\%SiC$, were tested and compared for their wear properties at different loads of 10, 15, and 20 N. From the results, it was found that reinforced particles had great influence on increasing the wear properties of the composite, whereas increased loading condition showed adverse effect on the wear properties.

Keywords Aluminum 7075 · Hybrid composite · Boron carbide · Silicon carbide · Pin-on-disc · Microstructure

---

S. K. Tiwari (✉) · S. Pal · A. Sharma · A. Dasgotra · J. K. Pandey (✉)
Department of Research and Development, University of Petroleum and Energy Studies, Dehradun, Uttarakhand 248007, India
e-mail: suniltiwari.me.utu@gmail.com

J. K. Pandey
e-mail: jkpandey@ddn.upes.ac.in

S. Pal
e-mail: spal@ddn.upes.ac.in

A. Sharma
e-mail: Abhishek_sharma@ddn.upes.ac.in

A. Dasgotra
e-mail: adasgotra@ddn.upes.ac.in

A. K. Dubey et al. (eds.), *Recent Trends in Industrial and Production Engineering*, Lecture Notes in Mechanical Engineering,
https://doi.org/10.1007/978-981-16-3135-1_14

1 Introduction

Aluminum and their alloys have been used in different manufacturing industries for a long time [1–4]. Their lightweight, low density, high strength, and corrosion resistance properties make them a suitable candidate for replacing iron and steel from many application areas [5]. The need of more enhanced properties in physical and chemical environments has led to the fabrication of aluminum composites and so on aluminum hybrid composites. Usually, the enhancement in properties of aluminum composites depends upon reinforced particles, fabrication processes, and optimized process parameters and techniques. In addition to this, strength can also be augmented by hardening and cryogenic treatments. There are several fabricating processes of aluminum composites including solid-state processing, liquid state processing, and in situ processes. Keeping in mind the cost effectiveness, many researchers have adopted casting processes, among which stir casting is the common and popular one [6–11]. Stir casting method for synthesizing aluminum composite needs certain considerations like, melting and pouring temperature, material of crucible stirrer and mold, RPM of stirrer, stirring time, stirrer design, preheating of mold, and variations in reinforcements. Many researchers have used different nanomaterials like carbon nanotubes, boron nitride nanotubes, graphene, nanofibers, etc., [5] for fabrication of aluminum composite. They have encountered the common problem of agglomeration of CNT and inhomogeneous mixing of other nanoparticles. The major problem with using these nanoparticles apart from agglomeration and inhomogeneous mixing is cost ineffectiveness. Many results have shown minimum increase in tribological properties of the aluminum composites reinforced with nanoparticles [12]. Selection of boron carbide as one of the reinforcements is worth because of its good hardness and shock absorption capacity, while its density near to that of aluminum helps in homogeneous mixing of reinforcement in the matrix material [1]. At the second hand, silicon carbide gives good harness and high thermal stability to the composite. Size of these reinforcing particles in micron makes them low-priced resulting the overall all synthesis of composite cost effective and also results in good enhancement in tribological properties of aluminum composites according to literature. Many researchers have fabricated aluminum hybrid composites through different channels and have so far found good results in terms of increased strength, hardness, and corrosion-resistant properties [6, 13, 14]. Shanmugaselvam et al. have fabricated hybrid aluminum composite using boron carbide, silicon carbide, and graphite as reinforcements via stir casting. They have concluded that, reinforcements had great influence over the composite in terms of increased properties in both the loading and unloading conditions. They have also ensured about the complete dispersion of the reinforcements in the matrix [14]. Hynes et al. have explained that, the presence of boron carbide in AA6060 composite had led to the improvement in bending strength, shear strength, and resistance to impact. They correspondingly explained that, by increasing the percentage of reinforcement in the matrix, hardness of the composite was increased [15]. Manikandan et al. have found considerable increase in tribological properties of Al-7075 composites reinforced with boron carbide and cow dung ash [14]. X. Canute and M. C. Majumdar have explained that the load applied on aluminum composite had the highest significance over the tribological properties. They found that, with increase in the load on the composite, wear resistance property was decreased [16]. A. Canakei and F. Arslan have also explained the positive effect of boron carbide in increasing the wear resistance properties of Al-2024 composite. In addition to this, they have also found that particle size of boron carbide has effective impact on the strength of the composite and proclaimed that, larger B_4C particle has shown good abrasive suspension as compared to smaller particles [17]. Pridhar et al. enlightened that higher B_4C content in the composite resulted in good wear properties [7]. Several authors have worked on fabricating and characterizing hybrid aluminum composite. This study illustrates synthesis of Al-7075 + B_4C + SiC composite via stir casting process and expounds the effect of reinforcing particles and load on tribological properties of the composites.

2 Materials and Method

In this study, to fabricate Al-7075 hybrid composite, boron carbide (B_4C) and silicon carbide (SiC) have been taken as reinforcing particles. Size of both the reinforcing particles is in the range of 40–50 μm. Four-blade stirrer of graphite was used for stirring the molten aluminum. Certain parameters considered prior to cast the composites have been mentioned in Table 1.

Al-7075 billets were kept in crucible and melted in electrical furnace. Once the melting of the aluminum was complete, it was initially stirred for 3–5 min to produce vertex inside it. Later, both the preheated reinforcements (B_4C and SiC) were put inside the vertex of molten aluminum formed inside the crucible simultaneously and stirred for next 8–10 min to ensure homogeneity of particles in the matrix. Finally, the stirred molten material was poured in the preheated mold in open environment and could solidify. Following the same step, three classes of composites were fabricated, namely $Al - 0wt\%B_4C + 0wt\%SiC$, $Al - 2.5wt\%B_4C + 2.5wt\%SiC$, and $Al - 5wt\%B_4C + 5wt\%SiC$. Samples for tribological investigations were prepared according to ASTM G99 standards. They were polished and cleaned with emery papers and acetone, respectively. Each class of composite sample was subjected to three category of loading conditions (10, 15, and 20 N) for wear test on pin-on-disc

Table 1 Process parameters of stir casting

Parameters	Values/specification/material
RPM of stirrer	500
Stirring time	8–10 min
Preheating of reinforcements	250 °C
Preheating of mold	200 °C

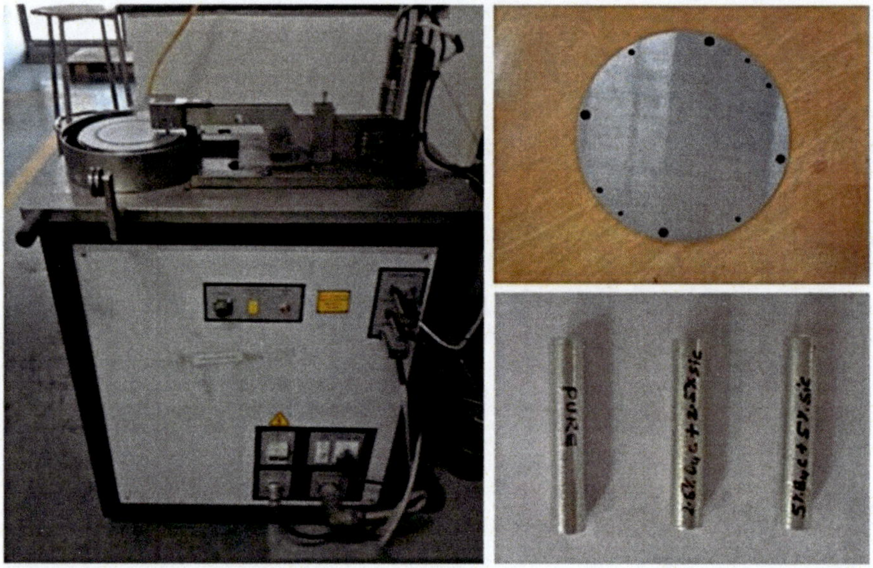

Fig. 1 Setup of pin-on-disc tribometer

tribometer. Test parameters considered for tribological investions were track diameter of 110 mm, load—10, 15, and 20 N, RPM—450, sliding time—15 min under dry sliding conditions and disc of high-speed steel. Figure 1 shows pin-on-disc tribometer setup, sliding disc, and test samples.

3 Results and Discussion

3.1 Wear Test Results

Results of the investigations have reflected that, there is appreciable effect of boron carbide and silicon carbide particles on enhancing the wear properties of the composites as shown in Table 2.

From the results, it can be clearly seen that, for every loading condition in wear test, Al $-$ 5wt%B_4C + 5wt%SiC has shown better result as compared to Al $-$ 2.5wt%B_4C + 2.5wt%SiC and Al $-$ 0wt%B_4C + 0wt%SiC. A sample result of wear test and frictional force has been shown in Fig. 2.

Table 2 Results of wear test

Load (N)	Composition/sample	Wear (μm)
10	$Al - 0wt\%B_4C + 0wt\%SiC$	278
	$Al - 2.5wt\%B_4C + 2.5wt\%SiC$	133
	$Al - 5wt\%B_4C + 5wt\%SiC$	100
15	$Al - 0wt\%B_4C + 0wt\%SiC$	342
	$Al - 2.5wt\%B_4C + 2.5wt\%SiC$	240
	$Al - 5wt\%B_4C + 5wt\%SiC$	236
20	$Al - 0wt\%B_4C + 0wt\%SiC$	411
	$Al - 2.5wt\%B_4C + 2.5wt\%SiC$	302
	$Al - 5wt\%B_4C + 5wt\%SiC$	302

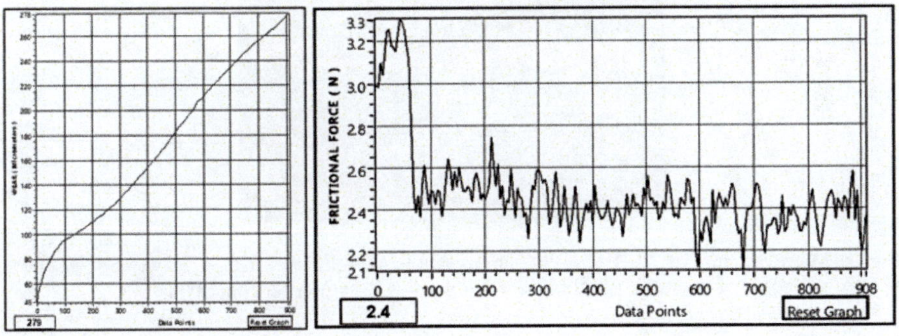

Fig. 2 Sample results of wear and frictional force

3.2 *Wear Results Discussion*

Results have shown that there is positive effect of reinforcing particles in enhancing the wear properties of the composite. $Al - 0wt\%B_4C + 0wt\%SiC$ has shown high wear as compared to $Al - 2.5wt\%B_4C + 2.5wt\%SiC$ and $Al - 5wt\%B_4C + 5wt\%SiC$ at 10 N of load. One of the reasons for this much of difference can be bad surface smoothening/flatness of the sliding disc, whereas presence of reinforcing particles in rest of the two composites as shown in Fig. 3 has shown good wear resistance properties among which $Al - 5wt\%B_4C + 5wt\%SiC$ has shown the best result at 10 N. Preheating of reinforced particles has helped in making good bond with aluminum matrix. Results also showed that with increase in percentage of reinforcement at 10 N, wear properties were increased. As the load was increased, pure sample, $Al - 0wt\%B_4C + 0wt\%SiC$ showed considerable increase in wear loss while rest of the two-composite showed higher wear loss as compared to the case in 10 N of load. This can be explained in the way that, in case of higher percentage of reinforcements and high magnitude of load, heat generation was more at the wear sample–disc interface. Due to the generation of high heat (high temperature), the

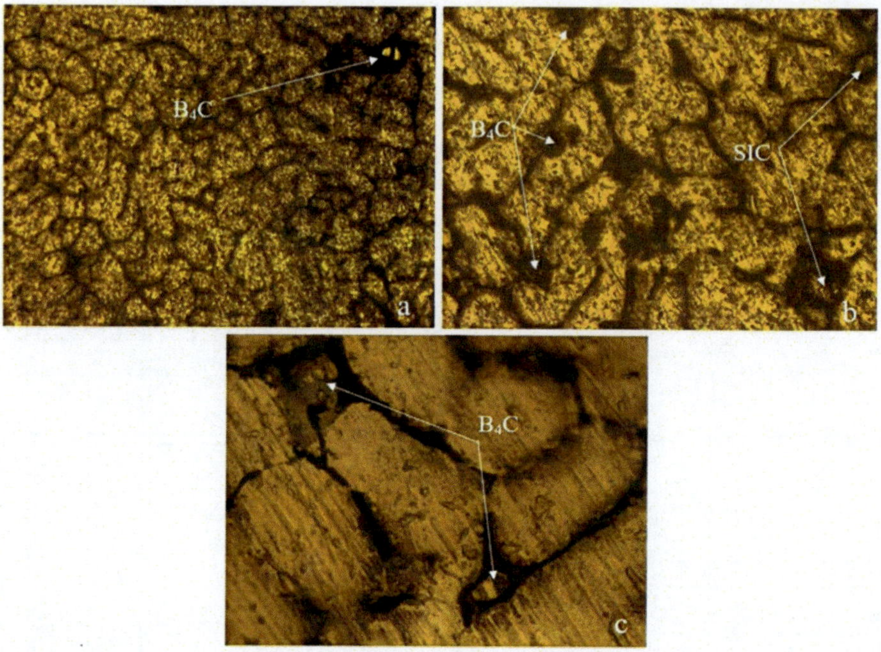

Fig. 3 Microstructure of Al-B_4C + SiC composite **a** Al − 2.5wt%B_4C + 2.5wt%SiC at 100×. **b** Al − 5wt%B_4C + 5wt%SiC at 200×. **c** Al − 5wt%B_4C + 5wt%SiC at 500×

material removal rate was increased as the sample–disc interface gets soften. Moreover, due to the high hardness of boron carbide and silicon carbide, the amount of aluminum matrix removed was more. This is because when reinforcing particles were detached from the matrix, they took away the large macroscopic part of the matrix with them due to the high heat generation and thus softening of the matrix. Figure 3 confirms that the damage of boron carbide particles was less but more cracks and pits were seen near grain boundaries. From the results, it can be proclaimed that Al − 5wt%B_4C + 5wt%SiC showed the best results at 10 N.

4 Conclusion

Different samples with different compositions of boron carbide and silicon carbide were fabricated through stir casting technique. Following conclusions were drawn from the experiments.

- With increase in weight percentage of reinforced particles, increase in wear resistance properties was observed.
- Al − 5wt%B_4C + %5wt%SiC showed the best composition among all the composites for different loading conditions.

- Al − 5wt%B_4C + %5wt%SiC showed the best result at 10 N of loading.
- Increase in load decreases the wear properties of the composites.

Acknowledgements Authors wish to acknowledge their sincere thanks to IIT Ropar for providing the experimental facility to fabricate the composite. Authors are thankful for the support and help from Central Instrumentation Centre, University of petroleum and Energy Studies, Dehradun, India and Department of Science and Technology, Ministry of Science and Technology, Government of India (DST/TM/WTI/2k16/245(G).

References

1. Hynes NRJ, Raja S, Tharmaraj R, Pruncu CI, Dispinar D (2020) Mechanical and tribological characteristics of boron carbide reinforcement of AA6061 matrix composite. J Brazilian Soc Mech Sci Eng 42(4):1–11. https://doi.org/10.1007/s40430-020-2237-2
2. Saleh B, Jiang J, Ma A, Song D, Yang D, Xu Q (2020) Review on the influence of different reinforcements on the microstructure and wear behavior of functionally graded aluminum matrix composites by centrifugal casting. Met Mater Int 26(7):933–960. https://doi.org/10.1007/s12540-019-00491-0
3. Gowrishankar MC, Hiremath P, Shettar M, Sharma S, Satish Rao U (2020) Experimental validity on the casting characteristics of stir cast aluminium composites. J Mater Res Technol 9(3):3340–3347. https://doi.org/10.1016/j.jmrt.2020.01.028
4. Shanmugaselvam P, Sasikumar R, Sivaraj S (2019) Investigation of hardness and tribological behaviour of aluminium alloy LM30 reinforced with silicon carbide, boron carbide and graphite. Springer Singapore
5. Tiwari SK, Singh H, Midathada A, Sharma S, Ravella UK (2018) Study of fabrication processes and properties of Al-CNT composites reinforced by carbon nano tubes—a review. Mater Today Proc 5(14):28262–28270. https://doi.org/10.1016/j.matpr.2018.10.109
6. Manikandan R, Arjunan TV, Akhil AR (2020) Studies on micro structural characteristics, mechanical and tribological behaviours of boron carbide and cow dung ash reinforced aluminium (Al 7075) hybrid metal matrix composite. Compos Part B Eng 183, no. Al 7075:107668. https://doi.org/10.1016/j.compositesb.2019.107668
7. Suresh V, Vikram P, Palanivel R, Laubscher RF (2018) Mechanical and wear behavior of LM25 aluminium matrix hybrid composite reinforced with boron carbide, graphite and iron oxide. Mater Today Proc 5(14):27852–27860. https://doi.org/10.1016/j.matpr.2018.10.023
8. Subramaniam B, Natarajan B, Kaliyaperumal B, Chelladurai SJS (2018) Investigation on mechanical properties of aluminium 7075—boron carbide—coconut shell fly ash reinforced hybrid metal matrix composites. China Foundry 15(6):449–456. https://doi.org/10.1007/s41230-018-8105-3
9. Pridhar T, Ravikumar K, Sureshbabu B, Srinivasan R, Sathishkumar B (2020) Multi objective optimization of wear characteristics on al 7075/al2o3/b4c composites-desirability approach. J Ceram Process Res 21(2):131–142. https://doi.org/10.36410/jcpr.2020.21.2.131
10. Kumaraswamy HS, Bharat V, Krishna Rao T (2018) Influence of mechanical & tribological behaviour of Al 2024 MMC fabricated by stir casting technique—a review. Mater Today Proc 5(5):11962–11970. https://doi.org/10.1016/j.matpr.2018.02.170
11. Kumar KS, Patnaik VS (2016) Experimental investigation on aluminium alloy composites for wear behaviour. Int Conf Electr Electron Optim Tech ICEEOT 2016:3846–3852. https://doi.org/10.1109/ICEEOT.2016.7755433
12. Pérez-Bustamante R et al (2012) Wear behavior in Al 2024-CNTs composites synthesized by mechanical alloying. Wear 292–293:169–175. https://doi.org/10.1016/j.wear.2012.05.016

13. Raja K, Chandra Sekar VS, Vignesh Kumar V, Ramkumar T, Ganeshan P (2020) Microstructure characterization and performance evaluation on AA7075 metal matrix composites using RSM technique. Arab J Sci Eng 0123456789. https://doi.org/10.1007/s13369-020-04752-8
14. Manikandan R, Arjunan TV (2020) Mechanical and tribological behaviours of aluminium hybrid composites reinforced by CDA-B4C. Mater Res Express 7(1). https://doi.org/10.1088/2053-1591/ab6b54
15. Hynes NRJ, Sankaranarayanan R, Tharmaraj R, Pruncu CI, Dispinar D (2019) A comparative study of the mechanical and tribological behaviours of different aluminium matrix–ceramic composites. J Brazilian Soc Mech Sci Eng 41(8):1–12. https://doi.org/10.1007/s40430-019-1831-7
16. Canute X, Majumder MC (2018) Mechanical and tribological behaviour of stir cast aluminium/boron carbide/fly ash composites. J Eng Sci Technol 13(3):755–777
17. Canakci A, Arslan F (2012) Abrasive wear behaviour of B4C particle reinforced Al2024 MMCs. Int J Adv Manuf Technol 63(5–8):785–795. https://doi.org/10.1007/s00170-012-3931-8

Investigations of Wear Behavior of Journal Bearing Materials

Priya Gajjal and Shekhar Gajjal

Abstract Behavior of wear in tribological condition in metal-based brass and graphite material at dry and wet sliding conditions. Wear test of these materials were performed on wear test machine. The test of wear in mass loss was performed under different loads 10, 20, 30 N with interval and at different 1000, 1200, 1400 rpm speed, etc. Wear characteristics of materials are investigated through load, speed, and time. The main objective is to investigate the wear rate of journal bearing materials at different loads and different speeds.

Keywords Wear · Wet and dry sliding · Brass · Graphite · Journal bearing

1 Introduction

Lubrication, wear, and friction are the major tribological principle subjects in the interacting surfaces. When two solid materials come in contact, relative motion between bodies occurs, the friction resistance opposes the motion at the contact surface, and wear mechanism is possibly to occur at mating parts. The characterization of the behavior of wear and friction processes at varying conditions shows the existence. They are highly considered as dependent on the scope and intricacy of the system, which is under investigation. In between two surfaces, lubricant is used in contact to lower the wear and friction. Researchers [1–4] show that coefficient of friction, wear loss is depending on pressure, load, speed, contact surface geometry, surface roughness, contact pairing material, systems dynamics, relative humidity, temperature, and lubrication. Many authors mentioned that wear characteristics and coefficient of friction of alloys and metals are dependent on the test conditions very strongly. In this connection, many researches have been done by using different test

P. Gajjal (✉)
AISSMS College of Engineering, Pune, India

S. Gajjal
Government Polytechnic Nasik, Nasik, India
e-mail: psgajjal@aissmscoe.com

A. K. Dubey et al. (eds.), *Recent Trends in Industrial and Production Engineering*, Lecture Notes in Mechanical Engineering,
https://doi.org/10.1007/978-981-16-3135-1_15

conditions, methods, and parameters for different material composites [5]. The friction and wear behavior regarding the load and the speed is found out. Unlu et al. [6] developed a journal bearing dedicated test rig to determine frictional coefficient in bronze composite material. Thus, here it is found that the friction force rises with increase in load and velocity. In the beginning, as dry friction occurs in the motion, frictional coefficient increases, and later, it declines. It is noticed that if the load increases, frictional coefficient decreases in dry condition. Zeren et al. [7] investigated that the Babbitt alloys, which is used widely now as sliding bearing material, show that after testing specimen in severe dry and wet conditions, in the occurrence of lubricating film, still dry sliding will appear from time to time between surface contact, generating local wear at mating surfaces. Local wear results in specimen material wearing out early before its life period. Anand and Achary [8] depict that gun metal gives excel performance compared to brass at dry and wet condition because of hardness variation. Brass exhibits increase in rate of wear. Junghans et al. [9] investigated that, at high speed, change in geometry of bearing of tribological property can affect coefficient of friction of specimen. Permeability may affect the modes of lubrication in this type of bearing material which may increase wear rate. Yunxin et al. [10] explained that in self-lubricating composite material, frictional coefficient and wear behavior of nickel/molybdenum disulfide changes its conformity in the presence of surface lubricating film formation. The probity of the lubrication film behaves like self-lubricating composite. In self-lubrication property, the optimal concentration interpreted in terms of theoretical as thin film solid lubrication. Embedded [11] bearings are having property of self-lubrication where base metal embedded in solid lubricants. SLEB (solid lubricants embedded) are using in many developed and developing countries. Embedded-based bearings may be used in different ways as specified; selection may be extensive in solid lubricant and metals. These bearings are used extensively in higher load, less speed, higher temperature, etc. Anti-corrosion property having graphite-additive lubricant is highly effective; endurance temperature is high, property of lamellar structure is good and very effective etc. Lamellar slippage enhances graphite and gives excellent lubricity at high loads [12, 13]. Oxide layer plays vital role at higher temperature, which gives good lubrication in graphite. The aluminum and graphite composites showed a low coefficient of friction and mass loss compared to base alloy in sliding dry condition and oil impregnate [14]. The increase in transition pressure seen at dry condition as graphite increases, which may be correlated to retain and formation of thick layer of solid lubrication film, which covers maximum composite area of surface in sliding. Thus, tribological film layer will minimize contact in surface metal, which shows minimum the frictional coefficient and reduces shear stress which may transmit to the total subsurface composite material [15, 16]. Few researchers investigated that PTFE and its matrix or composites having good capability to reduce wear, but at higher temperature, it will not able to sustain the wear loss, which in turn have adverse effect [17, 18]. Addition of particles of graphite which gives a thick layer on the surface in contact thus reduces wear. It has been depicted that the strength and wear property is better as compared to Al and its matrix [19–21].

2 Experiment Details

The main purpose of this paper is to investigate wear loss in journal bearing material of brass and graphite with dry bearing test rig, which is designed to investigate in situ condition. This test rig gives friction and wears study in situ condition, by using test bearing specimen mounted on the motor shaft which encloses in bush which is supported in machine shaft. Hence, performance and its results show very close to the true values and figures as in simulated condition in tribology, output gives friction and mass loss or rate of wear. Digitally, operating parameters and its yield value can be noted with control panel. Specifications of test rig is Motor-5HP, speed 3500 rpm, load cell is swinging field type, first beam is 150 kg, friction arm force is 'S' type with 500 kg, second radial arm load, the shaft is 50 mm ID, and digital control panel with rpm of shaft radially applied load with arm friction force.

During the experiments, specimen is fixed and tightened. The load is applied in running condition, when the shaft material and bearing material come in contact, friction and wear occurs.

Figures 1 and 2 show the dry bearing test rig in front view and side view. Figures 3 and 4 show the specimen material which is used for testing purpose.

In automobile sector, where the friction and wear of bearing is more concerned, graphite and brass material is used. It is having good strength, good wear resistance, and excellent mechanical properties. Mostly, these materials are used in bushes, bearings, pumps, valve and piston, gears, etc. Thus, in this paper, graphite and brass are used for study and investigation of tribological parameters as a test specimen, prepared by heat treatment.

The test was conducted using above test rig. Initially, the test specimen is fixed in position and rotating shaft is a cast iron, with small clearance that it will maintain lubricity in between the contacting parts. Then, as it starts the process and machine, load is applied and friction will be displayed and wear will be measured after and before test. For wet condition, SAE 40 is used as lubricant.

Fig. 1 Test rig: dry bearing machine

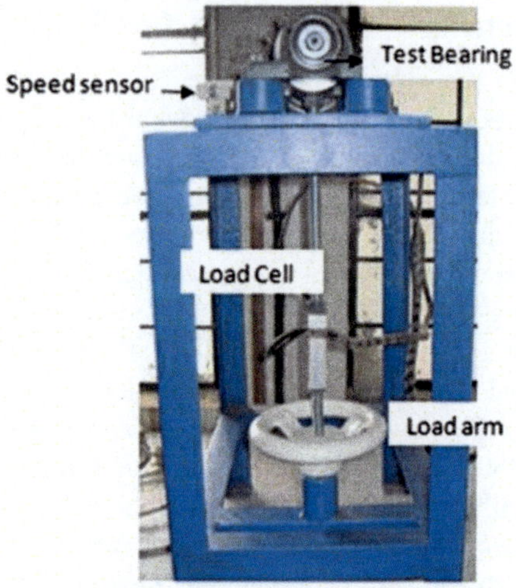

Fig. 2 Load cell attachment in dry bearing m/c

Fig. 3 Specimen of bearing material (Top view)

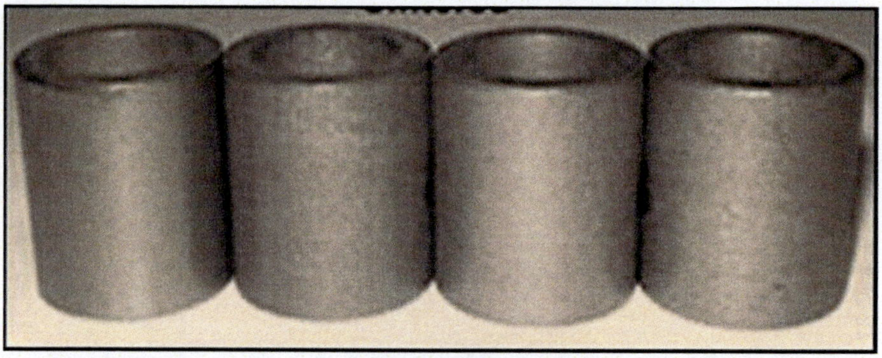

Fig. 4 Specimen of bearing material (vertical position)

3 Results and Discussions

The tests were conducted in wet and dry condition in dedicated test rig at speed 1000, 1200, and 1400 rpm and load 10, 20, and 30 N at an interval of time from 10 to 50 min. Wear is measured with mass loss measurement after and before test, and then, it is converted into wear rate. Wear rate is mass loss with distance in sliding condition. As shown in Figs. 3 and 4, the test specimens are used in brass and graphite material. The following graphs show the variation rate.

The variation in graph is plotted in wet and dry condition of test specimens as load with wear rate and time with wear rate. Figures 5, 6, 7, 8, 9, 10, 11, 12, 13, 14, and 15 depict that as time increases, wear rate decreases. It clearly shows that the strength and hardness reduces wear which minimizes wear, but at same time, if amount of graphite is more after particular amount, gives adverse effect which causes local wear in turn lubrication layer gets thinner and thinner. And Figs. 13, 14, 15 and 16 show that as load increases, the wear rate increases. In dry, the same is seen as the rate of wear is more compared to wet condition. This means that as speed increases, wear rate is also increasing but if we compare wet and dry condition, in dry wear is more.

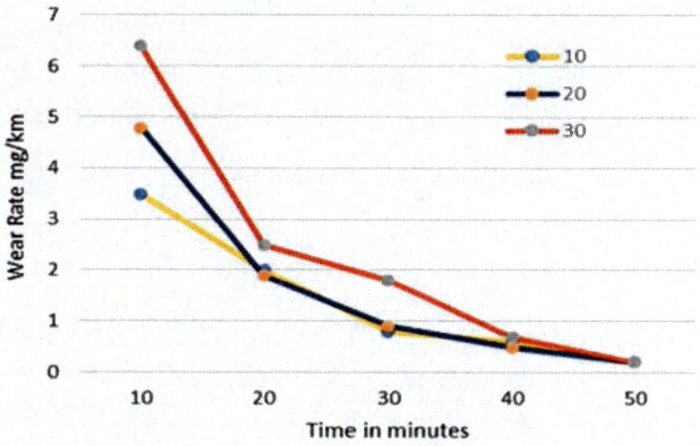

Fig. 5 Wear rate versus time at 1000 rpm brass in dry condition

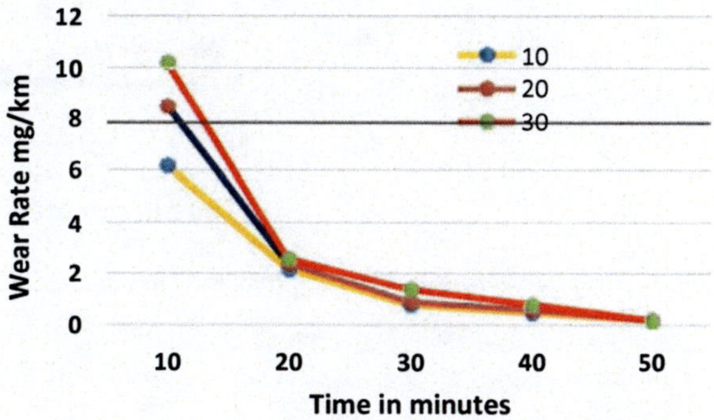

Fig. 6 Wear rate versus time at 1200 rpm brass in dry condition

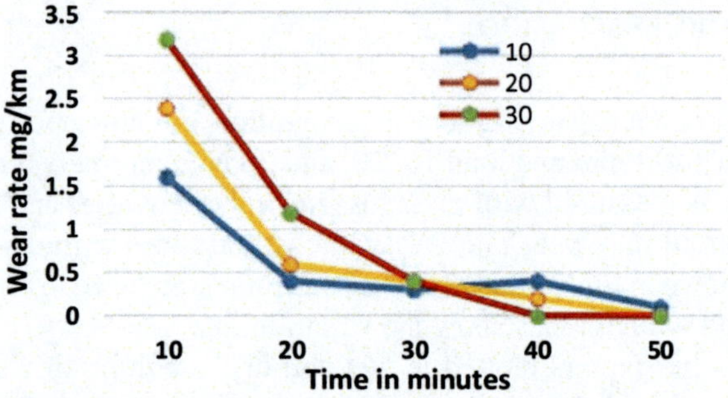

Fig. 7 Wear rate versus time at 1000 rpm graphite in dry condition

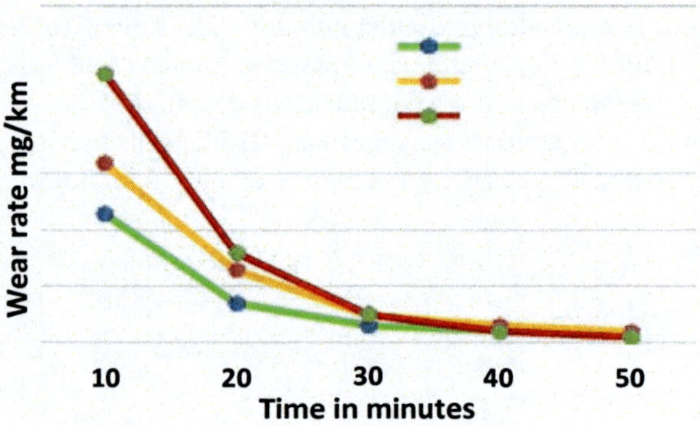

Fig. 8 Wear rate versus time at 1200 rpm graphite in dry condition

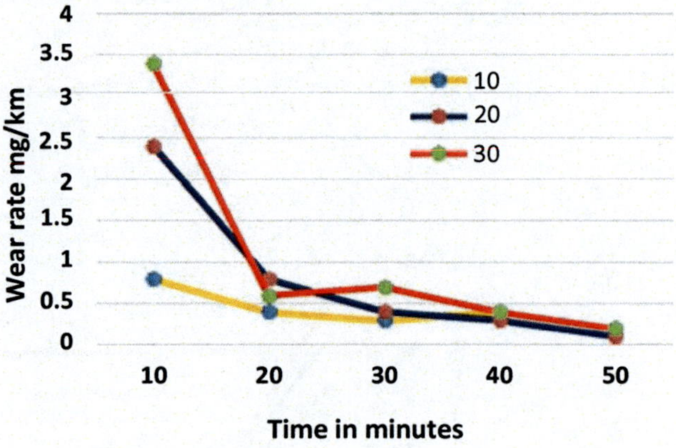

Fig. 9 Wear rate versus time at 1000 rpm brass in wet condition.

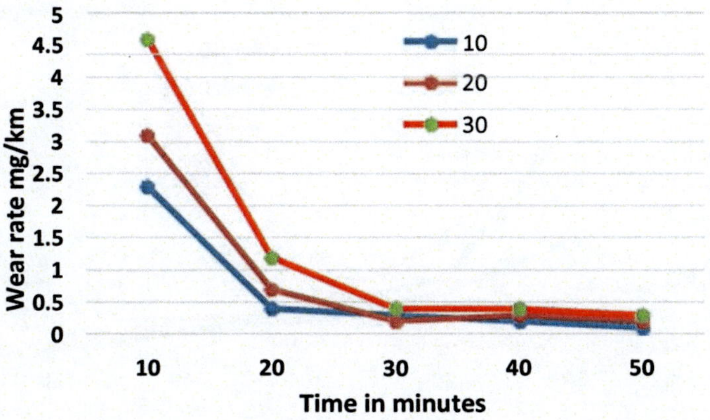

Fig. 10 Wear rate versus time at 1200 rpm brass in wet condition

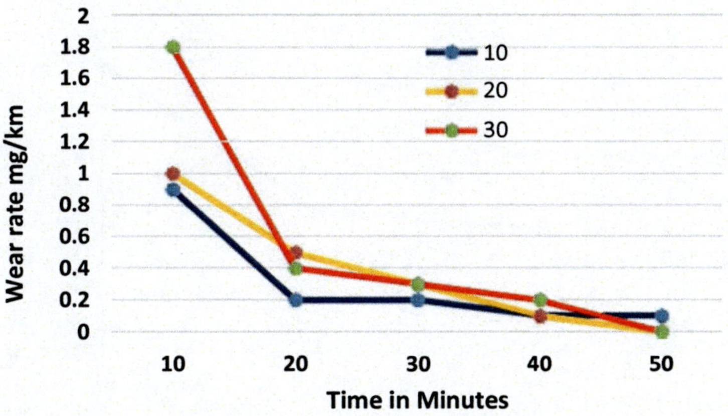

Fig. 11 Wear rate versus time at 1000 rpm graphite in wet condition

4 Conclusions

1. As test specimen with graphite shows less wear compared to brass due to its lubricity property and good strength, as graphite having high strength and after particular rubbing surfaces, its hardness will increase and wear rate is reduced.
2. The mass loss is maximum at high load in both the conditions, which depicts that at high load and speed, the lubrication film breaks due to high friction and temperature under contact area.
3. As speed increases, the temperature generation starts between the journal and bearing, causing local wear due to tribological conditions. As temperature increases, wear and friction also increases, lubrication film may reduce and its layer gets thin, thus, causing local wear.

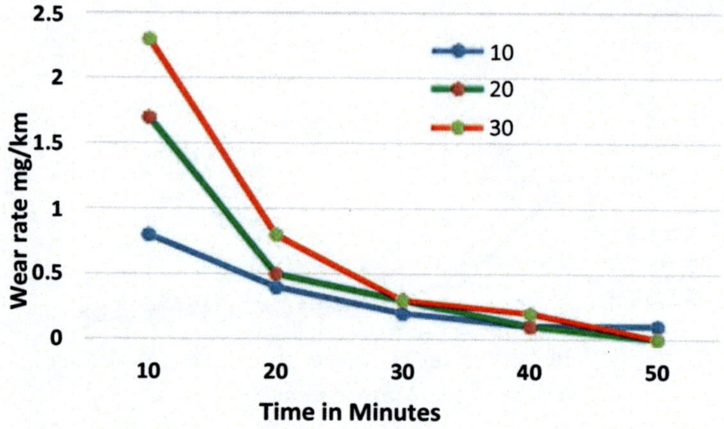

Fig. 12 Wear rate versus time at 1200 rpm graphite in wet condition

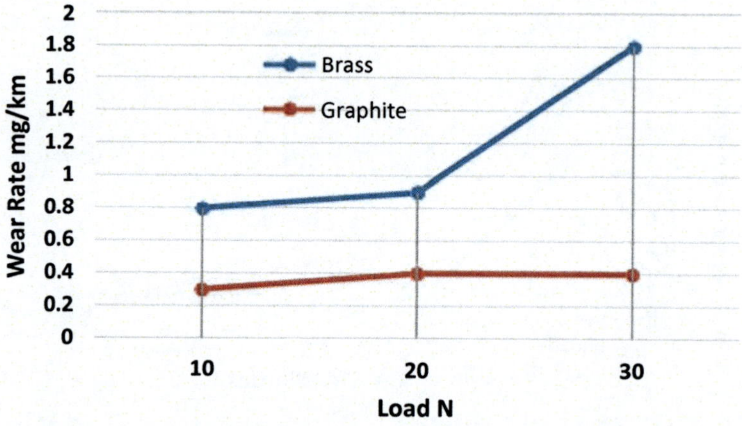

Fig. 13 Wear rate versus load at 1000 rpm (Brass and graphite in dry condition)

4. Compared to both materials, graphite shows high resistance to wear in both dry and wet condition, which shows in graph, as graphite is having high lubricity at higher temperature which generates a film layer between the surfaces in contact.

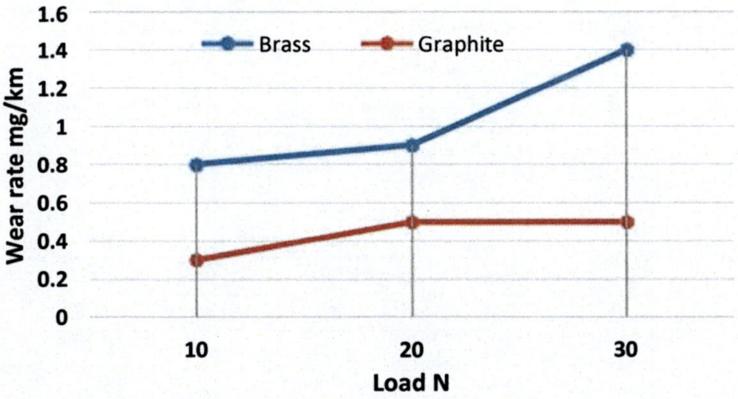

Fig. 14 Wear rate versus load at 1200 rpm (Brass and graphite in dry condition).

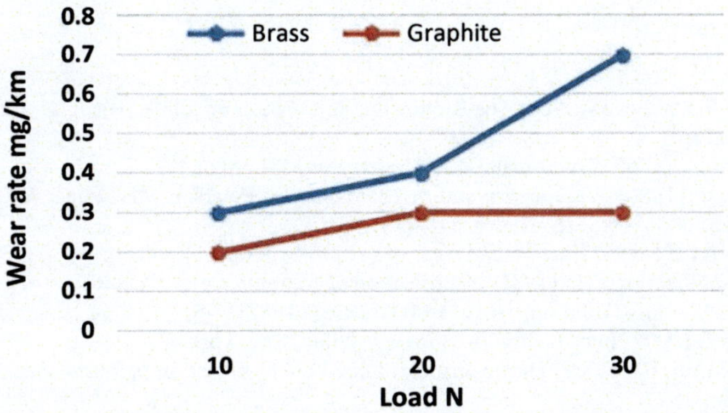

Fig. 15 Wear rate versus load at 1000 rpm (Brass and graphite in wet condition).

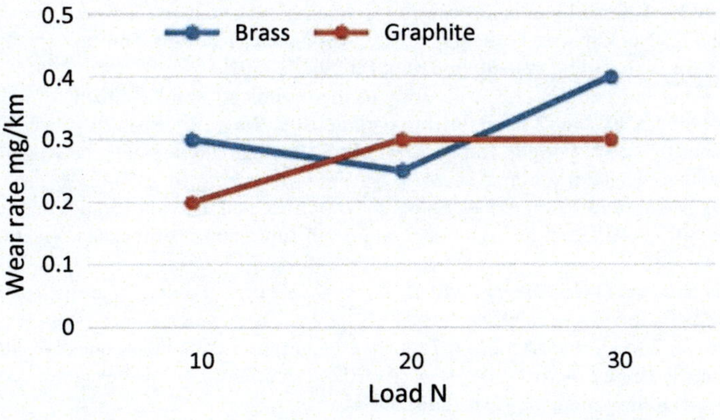

Fig. 16 Wear rate versus load at 1200 rpm (Brass and graphite in wet condition)

Table 1 Bearing material specimen under test

Parameters	Values (mm)
Inner diameter	40
Outer diameter	50
Width	10

Table 2 Chemical composition of bearing specimen

Alloys	Graphite (%)	Brass (%)
Cu	86	54
Zn	5.2	41.8
Sn	2.8	0.4
Pb	1.8	3.8
P	1.2	–

References

1. Bowden FD, Tabor D (1964) The friction and lubrication of solids, part II. Oxford Clarendon Press, Oxford
2. Kragelski IV (1965) Friction and wear. Butterworths, London
3. Archard JF (1980) Wear theory and mechanisms. In: Peterson MB, Winer WO (eds) Wear control handbook. ASME, New York, pp 35–80
4. Rabinowicz E (1995) Friction and wear of materials. Wiley, New York
5. Smith TF, Waterman MS (1981) Identification of common molecular subsequences. J Mol Biol 147:195–197. https://doi.org/10.1016/0022-2836(81)90087-5
6. Bhushan B (2002) Introduction to tribology. Wiley, New York
7. Sadik B, Unlu EA (2007) Determination of coefficient of friction in journal bearings. J Mater Des 28:973–977
8. Zeren A, Zeren, M, Feyzullahoglu E (2007) A study on tribological behavior of tin-based bearing material in dry sliding. J Mater Des (28): 318–323
9. Krishnan A, Manoharan Achary TS (2009) Wear behaviour investigations of journal bearing materials. In: 10th national conference on technological trends (NCTT09)
10. Junghans R, Neukirchner J, Schumann D, Lippman KH (1996) The use of sintered metal bearings for high sliding velocities. Tribol Int 29:181–192
11. Wu Y, Wang F, Cheng Y (1997) A study of the optimization mechanisms of solid lubricant concentration in Ni/MoS2 self-lubricating composite. Wear (205):64–70
12. Tongsheng L, Tao J, Cong P, Liu X (2001) Study on the tribological characteristics of solid lubricants embedded tin-bronze bearings. J Appl Polym Sci (80):2394–2399
13. Jinjun L, Shengrong Y, Jingbo W, Qunji X (2001) Mechanical and tribological properties of Ni-based alloy/CeF/graphite high temperature self-lubricating composites. Wear (249):1070–1076
14. Junan W, Herbert D (1998) Dry sliding wear behavior of molybdenum alloyed sintered steels. Wear 222:49–56
15. Akhlaghi F, Zare-Bidaki A (2009) Influence of graphite content on the dry sliding and oil impregnated sliding wear behavior of Al 2024—graphite composites produced by in situ powder metallurgy method. Wear (266):37–45
16. Amsallem C, Gaucher A, Guilhot G (1973) Unlubricated frictional behavior of sintered iron. Wear 23:97–112
17. Bhushan B (1999) Principles and applications of tribology. ISBN: 978-0-471-59407-9

18. Autay R, Missaoui S, Mars J, Dammak F (2019) Mechanical and tribological study of short glass fiber-reinforced PA 66. Polym Polym Composites 79–84
19. Swami MC, Dyad AA (2015) Comparative wear studies of three polytetrafluorethylene (PTFE) composites: a Taguchi approach. Int J Adv Res Eng Manage (IJAREM) 01(07)
20. Thareja P, Akhai S (2017) Processing parameters of powder, aluminum-fly ash P/M composites. J Adv Res Mater Sci Metall Eng
21. Shivdev S (2019) Wear behaviour of aluminium matrix hybrid composites—a review. J Adv Res Mech Eng Technol 6(2)

Metallurgical and Mechanical Behavior of Friction Stir Processed WE43/Nano-SiC Surface Composite

Shivali Singla, Amardeep Singh Kang, and T. S. Sidhu

Abstract In this paper, various FSP parameters have been made on the metal and mechanical properties of WE43 Mg alloy reinforced with nano-SiC particles. In addition, the geometric impact of different tools has been investigated in many advanced Mg alloys metal alloys using FSP. It has been found that the FSP of picnic geometry tools has significantly influenced aspects of integration. From 22.42 to 8.17 μm, the grain size of the al alloy dropped significantly. In addition, the square of geometric pin has shown excellent results in terms of metal and mechanical properties. Fine coefficient production is considered a comforting action produced by a square tool in a plastic environment that provides uniform distribution of particle strength within the metal matrix.

Keywords Friction stir processing · WE43 · SiC · Hardness

1 Introduction

Interest in magnesium alloys (Mg) has increased since the last decade due to its high properties, good texture, high corrosion resistance and durability [1]. The demand for the use of Mg alloys has increased in recent years such as building materials in the automotive industry, electrical products and car tires. [2]. To achieve this, various researchers have studied the effect of strengthening the ductile matrix with hard and heavy particles such as carbides, oxides, nitrides and borides [3, 4]. The addition of reinforcement has led to the integration of suitable properties such as low-temperature growth, certain high strength, good wear resistance, low weight, high thermal conductivity, durability and resistance to environmental corrosion [5].

S. Singla (✉)
IKG Punjab Technical University, Kapurthala 144603, India

A. S. Kang
Lovely Professional University, Phagwara 144411, India

T. S. Sidhu
Shaheed Bhagat Singh State Technical Campus, Ferozepur 152001, India

A. K. Dubey et al. (eds.), *Recent Trends in Industrial and Production Engineering*, Lecture Notes in Mechanical Engineering,
https://doi.org/10.1007/978-981-16-3135-1_16

Among the various methods of preparation for the manufacture of matrix, friction stir processing (FSP) metal has recently attracted the attention of many researchers before assembling Mg-matrix and nano-composites [6]. The FSP is based on the principles of friction stir welding (FSW) established at The Welding Institute (TWI) in the UK in 1991 as a solid form of joining process [7]. The FSP procedure was first proposed by Mishra et al. [8] and was originally developed for aluminum alloys. In FSP, the contraction temperature softens the matrix mixture, and the reinforcing particles are evenly distributed within the matrix by the stimulating action of the unused tool [8–10]. To date, extensive research has been done on FSP matrix nano-composites and composites, and the positive effects of FSP technology on grain refining and earth and mass material development have been demonstrated [11, 12]. Therefore, FSP is a method of grain refining and a very attractive process for compounding.

In this research work, the developed WE43/nano-SiC surface was developed by the FSP process. In addition, the effect of various FSP parameters was found. Also an analysis of the impact of the pin-pin geometry tool on the metallurgical and mechanical properties of the composite surface developed.

2 Experimentation

2.1 Materials

In the present study, we employed WE43 Mg alloy as a substrate. The detailed composition of the used substrate is given in Table 1. Specimens of 15–15 mm are prepared for the analysis of metal and mechanical properties. Specimens are polished to look like a face using various marks of emery paper and further polished with a diamond attach to the end of the mirror. After that, the specimens are cleaned with acetone to remove any foreign impurities from the polished specimens. Nano-Silicon Carbide (Nano-SiC) with more than 99% purity was purchased from Intelligent Materials Pvt Ltd, DeraBassi (India). The size of the Nano-SiC powder was in the range of 45–65 nm. Figure 1 shows FE-SEM morphology and EDS spectrum of nano-size SiC powder used.

Table 1 Elemental weight percentage (wt%) of WE43 Mg alloy

S. No.	Element	wt (%)
1	Yttrium (Y)	3.7–4.3
2	Rare earths (RE)	2.4–4.4
3	Zirconium (Zr)	0.4
4	Magnesium (Mg)	Balance

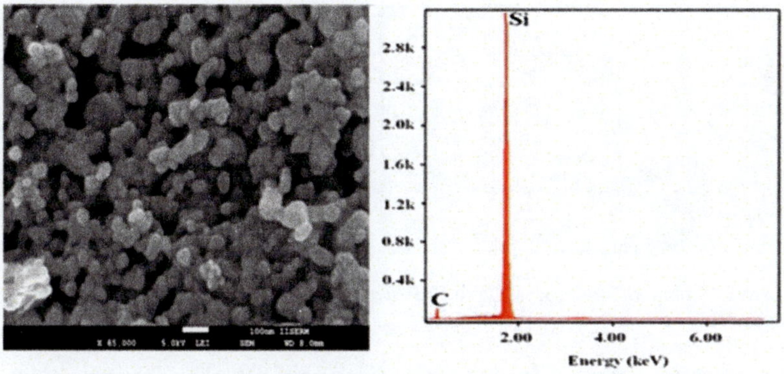

Fig. 1 Morphology and EDS spectrum of nano-sized SiC powder

2.2 FSP Setup Used in Study

The setting used for the processing analysis (FSP) is shown in Fig. 2. The thickness of the WE43 Mg alloy was 6 mm which was a rolled sheet. 2 mm wide holes are drilled in a straight line with a gap of 8 mm, and 2 rows of this adjustment are arranged in such a way that the holes of the opposing lines form a zigzag pattern with the first row of holes as shown in Fig. 3a. In these holes, nano-size SiC powder was filled. A straight cylindrical collision tool (SD 18 mm) without a pin was used to prevent the flow of reinforcement of the filling and their removal from the holes during the process after filling the pores in the holes. The sample used is shown in Fig. 3b. After that, different FSP pin tools (cylindrical, square and triangular taper) are used which are used to improve the WE43/nano-SiC earth integration. The moving angle of the tool was held at 2° and set constant. To process the rolled plates, various process parameters such as rotation speed (800 and 1700 rpm), cutting speed (30 and 60 mm/min) and geometric tools (cylindrical, square and triangular) were used. The effects of process parameters on various structures (nano-hardness and elastic modulus) and microstructure on process parameters are investigated.

Fig. 2 Modified vertical milling machine used as FSP setup

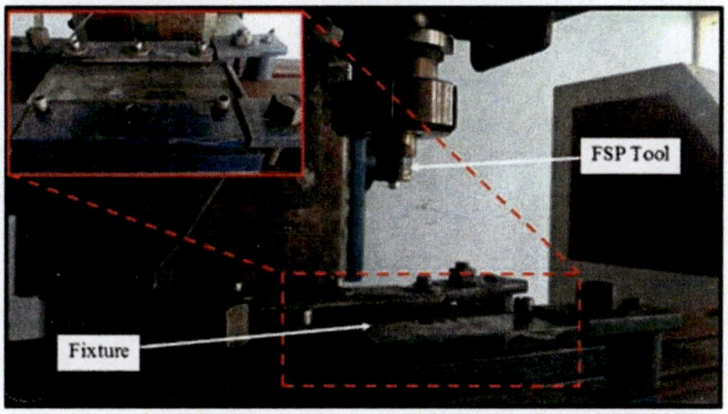

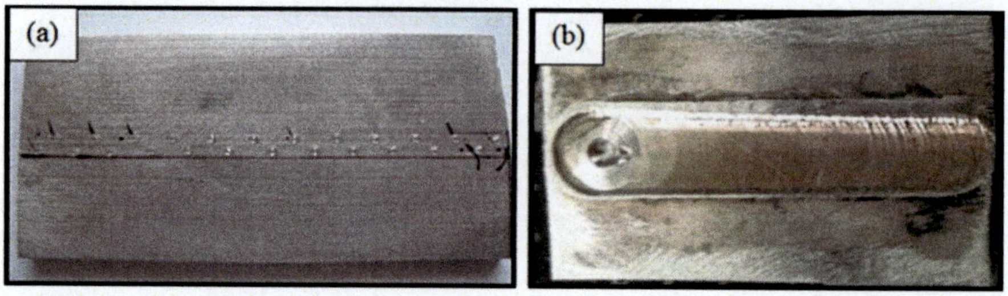

Fig. 3 **a** WE43 plate before processing. **b** FSPed sample

2.3 Microstructure Characterization and Phase Composition

Samples were prepared for microstructure analysis by a low-density diamond manufacturer from a region of 15 × 15 mm FSP and washed with ethanol. Subsequently, emery paper with various marks (400, 800, 1500, and 2000) was used to polish the specimens and then refined using a 0.1 and diamond paste and thus named 2% of Nital's blend. Subsequently, with the help of visual microscopy and electron microscopy, microstructure analysis was performed. Energy dispersive (EDS) spectroscopy combined with FE-SEM investigates the basic composition of WE43 Mg alloy before and after FSP. Electron backscatter diffraction (EBSD) map is designed for complete microstructure analysis and breast size research. The backscatter electron (BSE) detector was installed by FE-SEM. OIM analysis software was used to analyze data from EBSD results. Details of EBSD results were analyzed by OIM analysis software.

2.4 Mechanical Properties

Oliver-Pharr's method used with the help of a nano-indenter (Hyistron TI-950, Bruker's, Minneapolis, US), to measure the durability and hardness of WE43/SiC compounds using 1000 μN. By measuring the stiffness of the specimens, the nano-hardness of the specimens is guaranteed. The minimum hardness of the variety is rated with an indenting load of 0.49 N and a duration of 15 s with Vickers micro-hardness (HMV-G21ST, Japan). The sample cut-off surface was used for nano-induction testing and low hardness testing.

3 Result and Discussion

3.1 Metallurgical Behavior

A FE-SEM test for the empty and processed substrate is given in Fig. 4. Figure 4a, b shows the basic morphology of the inactive WE43 alloy. The EDS spectrum confirms the use of WE43 Mg alloy in the current study. The WE43/nano-SiC morphology shown in Fig. 5c, d reflects the morphology of the WE43 Mg alloy. It can be clearly seen from the FE-SEM micrograph that the nano-SiC is evenly distributed after the FSP in the WE43 alloy matrix. Nano-SiC was assembled and observed behind the FSP in the form of wide-diameter collections at a distance of 1–2 μm. This is because, at high rotation temperatures, the temperature generated in the processing area is high and causes cold cooling of SiC particles; thus, they combine and form clusters [13].

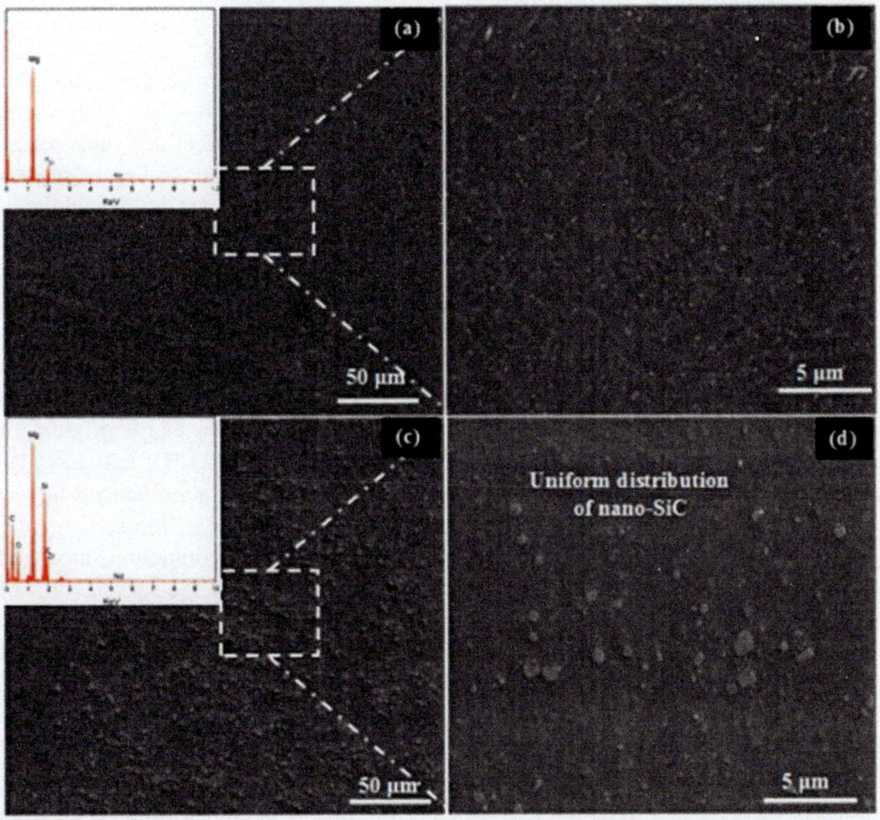

Fig. 4 FE-SEM of WE43 alloy **a**, **b** without FSP and **c**, **d** after FSP

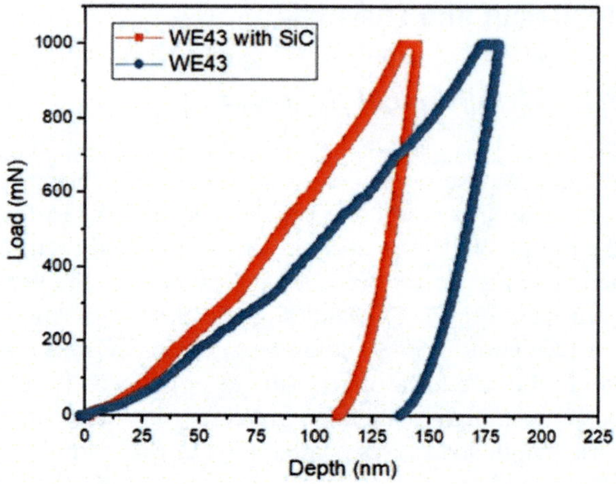

Fig. 5 Nano-indentation plots (p–h plots) of WE43 alloy before and after FSP

3.2 Mechanical Properties

A nano-induction graph was used to calculate the micro-hardness and elastic modulus of WE43/SiC components. Before and after the processing process, Fig. 5 shows the load associated with the depth of the WE43 alloy import structure. Compared to unprocessed WE43 composites, it can be shown that the penetration depth is lower in WE43/SiC composites, which is due to the fact that the minimum weight of WE43/SiC composites is much higher than that of unused WE43 composites. The grain particle size decreased and the combined strength increased as a result of the FSP process, which is resistant to penetration.

Observations also confirmed that WE43/SiC alloys have a much higher modulus compared to the pure WE43 alloy. Minor complications of WE43/nano-SiC composites, processed in the FSP method, have been reported. Figure 6 shows a mixture of unprocessed WE43 and micrographs WE43/SiC. It was found that a mixture of untreated WE43 and WE43/SiC had a minimum hardness of 90 HV and 150 HV, respectively. Notably, the penetration depth of the indenter was influenced by the presence of reinforcement which led to the development of less stiffness.

Figure 7 shows the micro-hardness line sites of WE43/SiC compounds arranged in various geometric metals. It was noted from this figure that the square form of pin-pin geometry (151.85 HV) encountered the greatest difficulty of the prepared compounds. However, the geometry of the triangular metal has also had an impact on the magnitude associated with small gravity, varying from about 140.85 HV. In the case of square metal geometry, the hardness value obtained was considered to be 129.26 HV. As noted, the geometry of the tool metal is important in the soil application of the substrate WE43 and, consequently, in the solidification of the solids. In the case of pin geometry, the FSP tool has been shown to generate high temperatures in the region of the production region and a large nugget related to the geometry of the square tool. In the case of the triangular pin geometry, however, it

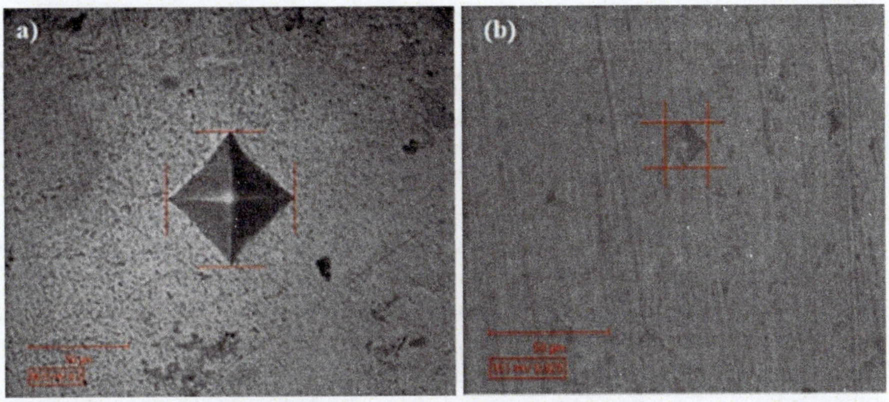

Fig. 6 Indentation image of WE43 alloy **a** before and **b** after FSP

Fig. 7 Variation of hardness of WE/SiC composite from center of FSPed zone prepared using different tools

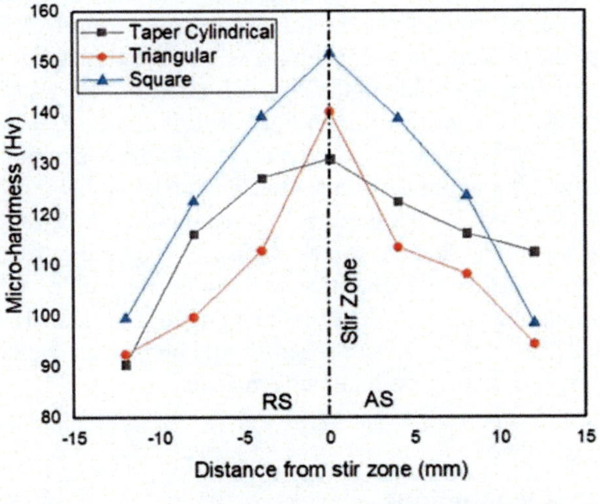

was observed that the welding area was larger than the geometry of the triangular and cylindrical pins. The geometry of the toolkit, as mentioned, is important in the ground analysis of the WE43 substrate and, consequently, the reinforcement. In the case of pin geometry asphalt, which is related to the geometry of a square tool, the FSP has also been shown to cause high temperatures of material in the manufacturing region and a large nugget. However, in the case of a triangular pin geometry pin, the welding area was considered larger than the geometry of the triangular pins and cylindrical pins. 122.58 HV hardness is produced at a distance of 4 mm from the center of the weld line, and the average hardness at a distance of 4 mm from the weld line is 113.85 and 138.58 HV, respectively, in the form of a pin and cylindrical pin geometry. The average hardness of WE43/SiC composites produced by cylindrical, square and triangular pin geometry 12 mm from the weld line and, however, is 112.45,

94.47 and 98.55 HV, respectively. Overall, pin-pin geometry was used to achieve the greater complexity of compounds.

It was noted that the micro-hardness and elastic modulus of the prepared specimens increased with a rotational speed from 800 to 1700 rpm. Typically, the rotation speed is related to the vibrating force applied to the operating environment by the rotating tool. Therefore, a large amount of heat was generated when the rotational speed was high. As a result, the gradient produced by the processing plant was high, resulting in a high degree of heat transfer, which resulted in the formation of small grains. Similarly, it can be found that the geometry of the speed of movement of the tool differs in proportion to the hardness of the micro-hard and elastic modulus. In terms of the results obtained from this analysis, it was found that the flow rate of the FSP tool is very important when it comes to processing time spent on the surface of the work by pin-pin geometry. The high flow rate of the FSP tool has hampered the full processing of the workspace which has not provided quality space as sufficient power can be accumulated for effective ground treatment. It therefore affects the small micro-load and elastic modulus of the collected work. Significantly, the observed value of the minimum hardness is approximately 15% with a travel speed of 30 mm/min rather than that of 60 mm/min. Similarly, in the case of the elastic modulus, the value obtained at 30 mm/min was found to be 53% greater than the value obtained at 60 mm/min. When it comes to tool pin geometry, square geometry has been shown to have a much higher degree of rigidity of 12% higher than a cylindrical pin geometry tool. The meaning behind the led patterns, previously, is well-tested. With regard to the modulus and the elasticity of the combination used, it was found that the values obtained by geometry by square tool tools were around 26.6 and 78.2% higher than the extended module obtained by the geometry of the triangular pin and cylindrical tool pin. In general, the pin geometry of the tools, rotation speed and shortcut speed should be kept as square, 1700 rpm and 30 mm/min, respectively, for optimal performance.

4 Conclusions

The following conclusions can be made from the present study:

- FSP tool pin profile substantially affects the metallurgical and mechanical properties of the surface composite. Square tool pin geometry has been found most suitable for effectively improving the metallurgical and mechanical properties.
- The generation of fine grains was attributed to the pulsating stirring action produced by square tool at the plasticized zone which gives the uniform dispersion of reinforcement particles within the metal matrix.
- Substantial reduction in grain size during the FSP of WE43 Mg alloy is resulted into significant improvement in micro-hardness of surface composite.

- The grain size of the processed compounds has decreased significantly, and as a result, the mechanical properties of the composite compounds have improved. It was found that from 22.42 to 8.17 μm grain size of WE43 alloy was significantly reduced.

Acknowledgements IKG Punjab Technical University, Kapurthala, India, has funded this work. To IKG Punjab Technical University, Kapurthala, India, the authors would like to express their sincere gratitude.

References

1. Polmear IJ (2005) Light alloys, 4th edn. Butterworth-Heinemann, Oxford, pp 29–96
2. Zettler R, Blanco AC, dos Santos JF, Marya S (2005) The effect of process parameters and tool geometry on thermal field development and weld formation in friction stir welding of the alloy AZ31 and AZ61. In: Neelameggham NR, Daplan HI, Powell BR (eds) Magnesium technology. TMS, San Francisco, pp 409–423
3. Suryanarayana C, Al-Aqeeli N (2013) Mechanically alloyed nanocomposites. Progress Mater Sci 58:383–502
4. Liu BZY, Xiao L, Wang WG, Ma ZY (2014) Analysis of carbon nanotube shortening and composite strengthening in carbon nanotube/aluminum composites fabricated by multi-pass friction stir processing. Carbon 69:264–274
5. Hsu CJ, Chang CY, Kao PW, Ho NJ, Chang CP (2006) Al–Al3Ti nanocomposites produced in situ by friction stir processing. ActaMaterialia 54:5241–5249
6. Qu J, Xu H, Feng Z, Frederick DA, An L, Heinrich H (2011) Improving the tribological characteristics of aluminum alloys by forming a nanocomposite surface layer using friction stir processing. Wear 271:1940–1945
7. Thomas WM et al (1991) Patent Application No. 9125978.8, December
8. Mishra RS, Ma ZY, Charit I (2003) Friction stir processing: a novel technique for fabrication of surface composite. Mater Sci Eng A 341(1–2):307–310
9. Bauri R, Yadav D, Suhas G (2011) Effect of friction stir processing (FSP) on microstructure and properties of Al–TiC in situ composite. Mater Sci Eng A 528:4732–4739
10. Dolatkhah A, Golbabaei P, BesharatiGivi MK, Molaiekiya F (2012) Investigating effects of process parameters on microstructural and mechanical properties of Al5052/SiC metal matrix composite fabricated via friction stir processing. Mater Des 37:458–464
11. Izadi H, Nolting A, Munro C, Bishop DP, Plucknett KP, Gerlich AP (2013) Friction stir processing of Al/SiC composites fabricated by powder metallurgy. J Mater Process Technol 213:1900–1907
12. Devaraju A, Kumar A, Kotiveerachari B (2013) Influence of rotational speed and reinforcement on wear and mechanical properties of aluminum hybrid composites via friction stir processing. Mater Des 45:576–585
13. Cao G, Zhang L, Zhang D, Liu Y, Gao J, Li W, Zheng Z (2019) Microstructure and properties of nano-hydroxyapatite reinforced WE43 alloy fabricated by friction stir processing. Materials 12(18):2994

Modal and Static Analysis for Analyzing the Effect of Loading on Crack Propagation Rate Using FEM

Sumit Shoor and Manpreet Singh

Abstract Modal analysis is an important aspect for determining the mode shape generated by a component when excited by vibration and considered as a powerful tool to find dynamic characteristics of a structure which further helps in strength and reliability in initial stage of designing. In real-life application mode shape plays a vital role for determining the displacement. In the current study, taper roller bearing was modeled as per manufacture catalog using Creo and file was exported to ANSYS. Modal analysis is being done to find out the natural frequencies and mode shapes. Static analysis is being done to study the effect of loading conditions on crack propagation It has been found from results that loading has significant effect on crack propagation rate. Hence, FEM provides reference for failure analysis, design, optimization, and saves material time and cost.

Keywords Taper roller bearing · Dynamic analysis · Mode shapes · Finite element methods

1 Introduction

Bearing is often subjected to dynamic conditions and used in rotating machine parts and more prone to failure. Monitoring of bearing is essential to minimize downtime and failure of machine [1]. Vibration occurs in any rotating machinery when body is subjected to different forces which further produces stress, strain, and noise. Vibration also produces different frequencies associated with different modes. Modes are known to be inherent properties of a structure. The structure dynamics is extensively analyzed by many researchers. In initial stage, behavior of structure can be analyzed and optimized. Dynamic behavior of machine tool are predicted by Zhang [2]. Different techniques used for modal analysis and crack propagation are signal processing [3–5] acoustic emission technology [6], statistical approach

S. Shoor · M. Singh (✉)
School of Mechanical Engineering, Lovely Professional University, Phagwara, Punjab 144411, India
e-mail: Manpreet.20360@lpu.co.in

A. K. Dubey et al. (eds.), *Recent Trends in Industrial and Production Engineering*, Lecture Notes in Mechanical Engineering,
https://doi.org/10.1007/978-981-16-3135-1_17

[7], FEM [8–10], and network methods [11]. ANSYS is a finite element software that includes preprocessing and post-processing using graphical user environment. In ANSYS, problem can be solved using batch mode or interactive mode. In batch mode, command line is used to create input. In Interactive mode, graphical user interface is used to perform operation from graphic window [12]. In this paper, GUI is used for modal analysis, and static analysis is carried out to study the effect on loading of crack propagation. Main components of taper roller bearing is inner race, outer race, cage, and taper rollers.

2 Mathematical Formulation

In any dynamic analysis stress–strain displacement, velocities and load are dependent upon time. Following steps were used for carrying out dynamic analysis of bearing by using time frame model: (1) Creating Finite elements, (2) Assuming displacement model

$$U(x, y, z, t) = [N(x, y, z)]\, Q(t) \tag{1}$$

2.1 Deriving Element Stiffness Matrices and Load Vector

2.1.1 Expression for Stress and Strain Relations

$$\varepsilon = [B]Q : \quad \sigma = [D][B][Q] \tag{2}$$

$$M^{e} = \text{Element mass matrix} = \iiint_{V(e)} \rho [N]^{T}\, [N]\, DV \tag{3}$$

$$K^{e} = \text{Element Stiffness matrix} = \iiint_{V(e)} [B]^{T}\, [D]\, [B] DV \tag{4}$$

$$C^{e} = \text{Element Damping matrix} = \iiint_{V(e)} \mu [N]^{T}\, [N]\, DV \tag{5}$$

$$\text{Surface Forces} = \iiint_{V(e)} [N]^{T}\, \varphi Ds1 \tag{6}$$

$$\text{Body Forces} = \iiint_{V(e)} [N]^{T}\, \varphi Dv \tag{7}$$

2.2 *Solving the Equation and Applying Boundary Conditions or Initial Conditions*

2.3 *Assemble the Element Matrices and Vectors*

$$[M]\ \ddot{Q}(t) + [C]\ \dot{Q}(t) + [k]\ Q\ (t) = P(t) \tag{8}$$

where $Q =$ Vector of nodal acceleration in the global system.

3 Finite Element Analysis

3.1 *Defining the Problem: Pre-processing*

(1) Creating the three-dimensional model, (2) Selection of element type, (3) Defining material properties, (4) Meshing, (5) Creating contact between taper roller and inner race, taper roller and outer race, (6) Applying loads and boundary conditions, (7) Post-processing [13].

3.2 *Creating Three Dimensional Model*

Three dimensional model was built using Creo as shown in Fig. 1 and exported in ANSYS for analysis. Taper roller bearing mainly composed of outer race, inner race, roller, and cage. Figure 1 shows the dimensional CAD model and meshed model of taper roller bearing [14].

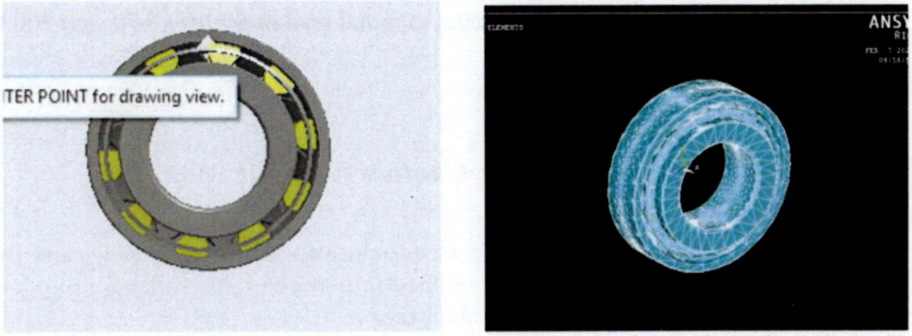

Fig. 1 Outer race defect and meshed model

3.3 Selection of Element Type

Element selected for analysis of bearing is Brick 8 node 185. Element used for three-dimensional modeling element is having characteristics of eight nodes and three DOF at each node [15].

3.4 Defining Material Properties

It is assumed that taper roller bearing material is isotropic with having following material properties: density: 7750 kg/m^3, Young's modulus: 200,000 Mpa, Poisson's Ratio: 0.29.

3.5 Meshing

Discretization of body into small number of elements is known as meshing. Two types of meshing were mostly used—free meshing or a mapped meshing. Free mesh has no particular pattern and not having any restriction to element shape. Compared to a free mesh, a mapped mesh is having a definite element shape and the pattern of the mesh. In this analysis, free meshing is used for analysis of taper roller bearing.

3.6 Loads/boundary Conditions

Loads/boundary conditions are applied to ANSYS to obtain a solution, e.g., degree of freedom is applied to restrict the movement of body in space which is known as boundary conditions. In this static analysis, outer race completely constrained in all degree of freedom and bearing is subjected to radial load of 50, 100, 150, and 200 N Fig. 2.

3.7 Define Contact and Check Contact Element

The multi-body contact phenomena are used to perform the contact analysis of the taper roller bearing. Contact pair is developed in between taper roller bearing and outer race and taper roller bearing and inner race.

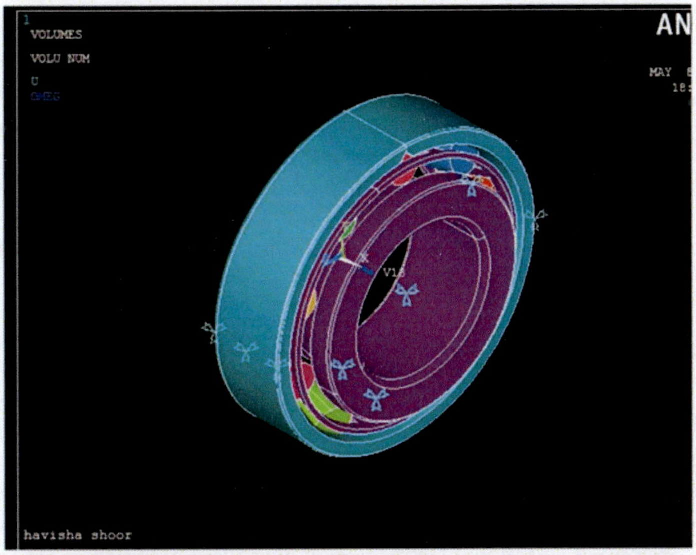

Fig. 2 Boundary conditions

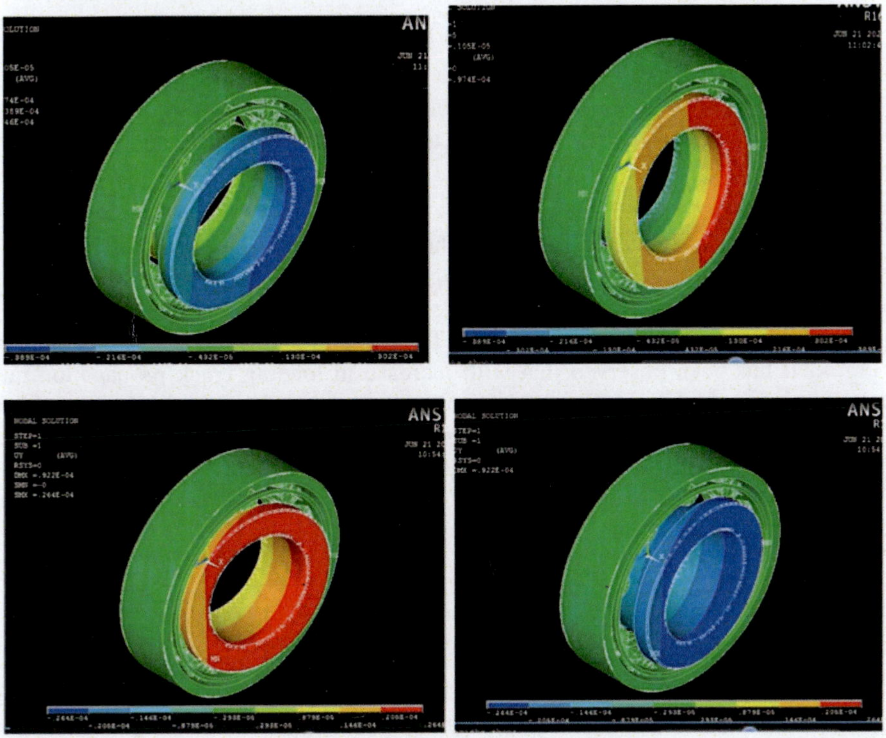

Fig. 3 Mode shape of defective bearing

3.8 Post Processing

Once the solution has been calculated, the ANSYS postprocessors are being used to review the results. Post processing is having different subparts, e.g., deformed shape, stresses, nodal temperature, list of all values, etc.

4 Result and Discussion

4.1 Mode Shape Analysis

Mode shape are the essential properties of any part or structure and greatly influenced by property of structure [14]. Mode shape will change if mass damping stiffness or boundary condition changes. Mode shape of taper roller bearing is simulated using FEM and defect was created on outer race as shown in Fig. 3. Angular velocity of 150 rad/s is applied and modal frequencies which were obtained in Hz are 50, 200, and 452 Hz for taper roller bearing. The frequency 50 Hz is the torsional bending and 200 Hz is the longitudinal bending and 452 Hz is the transverse bending obtained from simulated results.

4.2 Loading Effect on Crack Propagation

Effect of loading on crack propagation is shown in Fig. 4. Radial load was applied and was gradually increased to analyze the effect on crack propagation.

It has been observed that crack propagates faster as load increases. For plotting the graph, two nodes are selected at crack tip and graph is plotted. Effect of loading on crack propagation rate is shown in Table 1 and simulated result is shown in Fig. 4. It has been observed that with increase in loading crack propagation rate also increases; it can be witnessed from Table 1. It has been found that at load of 50–200 N, crack propagation rate increases 74.3%. Crack propagation rate increases 300 times approximately by increasing load from 50 N to 200 N.

5 Conclusion

Dynamic characteristics play a crucial role in design of machine system. Modal testing can be considered as reliable tool for testing the dynamic behavior of rotating machinery. Dynamic analysis is done using FEM and different frequencies were evaluated corresponding to different mode shapes which further helps to determine critical part of bearing during resonance. Different frequencies were extracted using

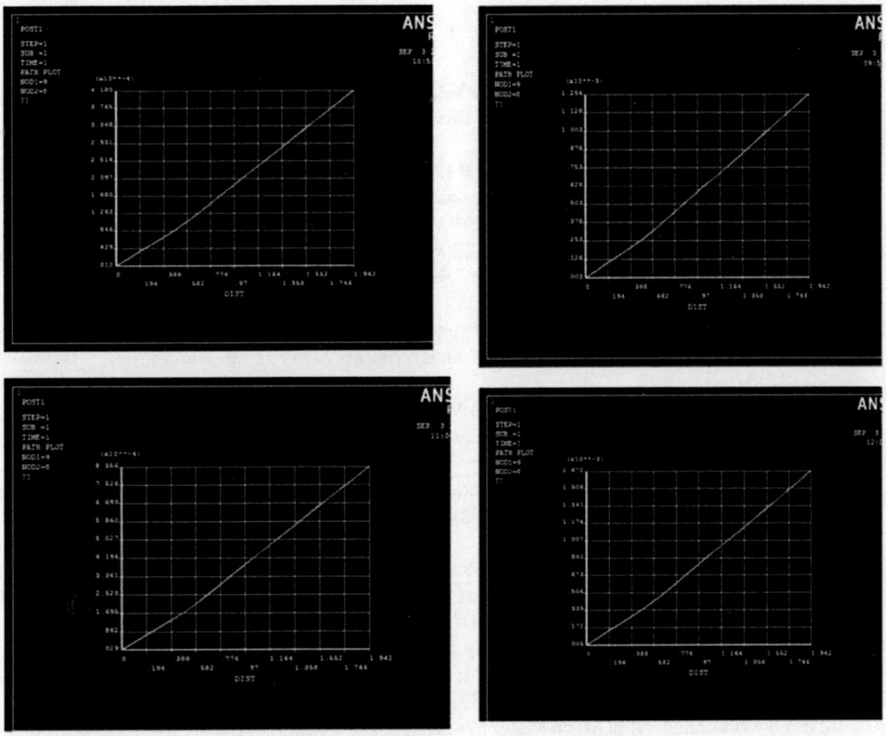

Fig. 4 Effect of loading on crack propagation

Table 1 Loading effect on crack propagation

S. No.	Load (N)	Crack propagation (mm)
1	50	0.41
2	100	0.83
3	150	1.2
4	200	1.6

block Lanczos method. Mode shape generated from modal analysis has different characteristics like torsional bending and combination of both. Further static analysis is also carried out to obtain crack propagation rate for different loading conditions. It is revealed that loading conditions directly influence the crack propagation rate to great significance as loading increases, crack propagation also increases.

References

1. Minhas N, Nikhil N, Banwait SS (2019) Vibration analysis and fault identifications of rolling element bearings-a review. Int J Mech Prod Eng Res Dev 9:1133–1142. https://doi.org/10.24247/ijmperdaug2019117
2. Zhang GP, Huang YM, Shi WH, Fu WP (2003) Predicting dynamic behaviours of a whole machine tool structure based on computer-aided engineering. Int J Mach Tools Manuf 43:699–706. https://doi.org/10.1016/S0890-6955(03)00026-9
3. Ebersbach S, Peng Z, Kessissoglou NJ (2006) The investigation of the condition and faults of a spur gearbox using vibration and wear debris analysis techniques. Wear 260:16–24. https://doi.org/10.1016/j.wear.2004.12.028
4. Adewusi SA, Al-Bedoor BO (2001) Wavelet analysis of vibration signals of an overhang rotor with a propagating transverse crack. J Sound Vib 246:777–793. https://doi.org/10.1006/jsvi.2000.3611
5. Al-Badour F, Sunar M, Cheded L (2011) Vibration analysis of rotating machinery using time-frequency analysis and wavelet techniques. Mech Syst Signal Process 25:2083–2101. https://doi.org/10.1016/j.ymssp.2011.01.017
6. Choudhury A, Tandon N (2000) Application of acoustic emission technique for the detection of defects in rolling element bearings. Tribol Int 33:39–45. https://doi.org/10.1016/S0301-679X(00)00012-8
7. Igba J, Alemzadeh K, Durugbo C, Eiriksson ET (2016) Analysing RMS and peak values of vibration signals for condition monitoring of wind turbine gearboxes. Renew Energy 91:90–106. https://doi.org/10.1016/j.renene.2016.01.006
8. Yongqi Z, Qingchang T, Kuo Z, Jiangang L (2012) Analysis of stress and strain of the rolling bearing by FEA method. Phys Procedia 24:19–24. https://doi.org/10.1016/j.phpro.2012.02.004
9. Xin HR, Zhu L (2014) Contact stress FEM analysis of deep groove ball bearing based on ANSYS workbench. Appl Mech Mater 574:21–26. https://doi.org/10.4028/www.scientific.net/AMM.574.21
10. Lostado R, Martinez RF, Mac Donald BJ (2015) Determination of the contact stresses in double-row tapered roller bearings using the finite element method, experimental analysis and analytical models. J Mech Sci Technol 29:4645–4656. https://doi.org/10.1007/s12206-015-1010-4
11. Huang R, Xi L, Li X et al (2007) Residual life predictions for ball bearings based on self-organizing map and back propagation neural network methods. Mech Syst Signal Process 21:193–207. https://doi.org/10.1016/j.ymssp.2005.11.008
12. Gokhale NS, Deshpande SS, Bedekar SV, Thite AN (2008) Practical finite element analysis, 2nd edn, p 445
13. Singeresu SR (2011) The finite element method in engineering (Isbn: 9781856176613)
14. Kathryn TM (2017) ANSYS mechanical APDL for finite element analysis
15. Nguyen-Schäfer H (2019) Computational tapered and cylinder roller bearings

Optimization and Validation of Material Removal Rate During EDM of HSLA Steel

Kulwinder Singh, Rakesh Goyal, Jaswinder Singh, Rupesh Gupta, and Manuj Madan

Abstract In this experimental work, electrical discharge machining (EDM) process is employed to machine the high-strength low-alloy (HSLA) steel. This work proposed the effect of machining input parameters such as current (A), pulse on time (B), and pulse off time (C) on material removal rate (MRR). As per Taguchi approach, fractional factorial technique is adopted during design of experiment (DOE) to reduce the time and cost of experimentation, so orthogonal array (L_9) is selected. The output is converted into signal-to-noise ratio (higher is better) for MRR. Analysis of variance (ANOVA) is performed to find the significance level of various input parameters at 95% confidence level. Optimum level of input parameters is determined with analysis of means (ANOM). Current is found the most significant parameter to improve MRR. The confirmation result shows the improvement of 17.78 mm^3/min in MRR at optimum settings.

Keywords EDM · HSLA steel · Taguchi · ANOVA · ANOM · Material removal rate

1 Introduction

Advancement of material science produces new materials that are widely used in aerospace, missile, surgical, tool and dies, defense industry, etc. [1]. New material such as super alloy, composite, ceramics, and different grades of steel required special process of machining to meet the demand of good surface finish and complex geometry of component [2, 3]. Conventional processes such as milling, turning, drilling, and shaping cannot be performed on harden material with complex geometry. Non-conventional machining processes are come into existence to overcome

K. Singh (✉) · R. Goyal · J. Singh · R. Gupta
Chitkara University Institute of Engineering and Technology, Chitkara University, Punjab, India
e-mail: kulwinder.singh@chitkara.edu.in

M. Madan
Chitkara Business School, Chitkara University, Punjab, India

A. K. Dubey et al. (eds.), *Recent Trends in Industrial and Production Engineering*, Lecture Notes in Mechanical Engineering,
https://doi.org/10.1007/978-981-16-3135-1_18

the difficulties occurred during machining with conventional process of hard and brittle material, supper alloy, tungsten, Inconel 718, hardened steel, high-strength low-alloy (HSLA) steel, and ceramics materials [4, 5]. HSLA steel is also known as micro-alloyed steel. It has low carbon percentage than conventional steels. HSLA steel contains addition of small alloying elements in steel such as vanadium, titanium, niobium, and aluminum. Addition of alloying elements less than 0.1% is desirable to enhance the strength and hardness. HSLA steel provides great opportunity in the automobile sector to reduce the weight and maintaining strength for given design load conditions. HSLA steel is mainly used in road vehicles, construction machinery, heavy duty vehicles, oil pipelines, defense equipment, bridges, power transmission equipment, etc.

HSLA steel selection is due to its better mechanical property, high strength-to-wear ratio, high corrosion resistance, and formability. Such properties of HSLA steel resist the machining with conventional machining process. To overcome this difficulty, new developed advanced machining process such as EDM, water jet machining, and abrasive jet machining can used for HSLA steel. EDM is selected due to its advantage over other non-conventional machining process. It is physical contact-free machining that totally eliminates the chances of stress development on work piece.

The important parameter of machined surface, surface integrity, significantly affects its performance such as creep, micro cracks, surface unevenness, recast layer, heat-affected zone, and corrosion. In past, several investigations have been carried out for machining of HSLA steel using different machining processes. The effect of EDM parameters such as current, voltage, rotation of electrode, flushing pressure, and powder-mixed dielectric on the responses viz., material removal rate, surface roughness, electrode wear rate, and micro hardness were analyzed in past years [6–8].

Taguchi proposed signal-to-noise ratio (S/N ratio) as an output value of performance characteristics. Signal-to-noise ratio is the conversion of several repetitions into one value. It measures the variability in the output due to noise factors. As per Taguchi, it is the quality loss if any value is deviating from the target. Generally, average value of the repetitions was considered to measure the data which represents only accuracy of process. But S/N ratio represents both accuracy and precision of the process [9–12].

In this present work machining of HSLA steel is carried out using the electrical discharge machining. The effect of machining parameters on material removal rate (MRR) is evaluated and optimized using Taguchi approach.

2 Material and Methods

2.1 Work Material and Experimental Conditions

The experimental work was performed on Electronica die sinking EDM (EMS 5535). The electrode and workpiece setup on EDM is shown in Fig. 1. HSLA steel work piece was machined with cylindrical shape copper electrode of size Ø16 mm × 100 mm. Positive polarity of electrode was selected to conduct the experimental work on HSLA steel specimens of size 100 mm × 50 mm × 5 mm. Commercial grade spark erosion oil (SEO 450) was used as dielectric fluid for flushing purpose during machining. In this experimental study, current along with pulse on and off time is selected as process parameters to evaluate the as material removal rate (MRR). Experimental machining conditions of work piece, electrode material, and machine are given in Table 1. Side flushing method is used to clean the debris from working zone. Machining time of 25 min is selected to minimize the measurement error. Table 2 is representing the detail of input parameters and their levels.

Fig. 1 Tool and work piece setup on EDM machine

Table 1 Experimental machining conditions

Machining conditions	
Machine used	Die sink EDM (EMS5535)
Electrode	Copper
Workpiece	HSLA steel
Dielectric	EDM oil SEO 450
Flushing condition	Side flushing
Test time	25 min
Polarity	+ve electrode, –ve workpiece

Table 2 Detail of input parameters

Parameters	Symbol	Notation	Level 1	Level 2	Level 3
Current	I	A	10 A	18 A	26 A
Pulse on time	T_{on}	B	11 μs	55 μs	95 μs
Pulse off time	T_{off}	C	5 μs	7 μs	9 μs
Voltage	V	50 V			
Flushing Pressure (F_p)	FL	0.50 kg/cm^2			
Polarity	PL	+ve electrode			

Generally, non-conventional machining processes have low MRR comparative to conventional machining process. So MRR is considered as one of the important response during the optimization of non-conventional machining processes.

In this study, MRR is determined by using Eq. 1.

$$\text{MRR} = \frac{W_b - W_a}{\rho * t} \tag{1}$$

where

W_b = Weight of specimen before machining

W_a = Weight of specimen after machining

ρ = Material density

t = Machining time.

To improve the performance characteristics, the optimization of parameter design is necessary. A classical design parameter is used to carry out the large number of experiments. However, as per Taguchi, a fractional factorial orthogonal array is used for experimental work for three parameters, each having three levels. Based on Taguchi proposed method, the data is transformed and converted into its variability. Such variation is called signal-to-noise (S/N) ratio. It consolidates several repeated value in one and check the variability present in experimental result. Loss function in terms of signal-to-noise ratio is to be used to check the deviation between estimated and experimental value. Generally, the signal-to-noise (S/N) ratio is classified into three categories: Lower is better (S/N_{LB}), nominal is better, (S/N_{NB}) and higher is better (S/N_{HB}). Signal-to-noise (S/N) ratio helps to predict the combination of optimal parameter. Finally, a confirmation test was conducted to verify the predicted and experimental value at optimal setting of parameters.

As MRR (higher is better quality characteristic) is required high, so S/N_{HB} is calculated by using following Eq. 2:

$$\text{S/NHB} = -10\log\left(\frac{1}{r}\sum\nolimits_{i=1}^{r}\frac{1}{y_i^2}\right) \tag{2}$$

Table 3 Experimental design with results

Trial no	Input parameters			Avg MRR	MRR S/N ratio
	A	B	C		
1	10	11	5	1.0657	0.552
2	10	55	7	1.4513	3.235
3	10	95	9	1.6167	4.172
4	18	11	7	8.5487	18.638
5	18	55	9	10.266	20.228
6	18	95	5	12.04	21.612
7	26	11	9	17.521	24.871
8	26	55	5	18.322	25.259
9	26	95	7	19.979	26.011

3 Design of Experiments

A fractional factorial orthogonal array of $L_9$is used to conduct the experimental for three factors (current, pulse on time, and pulse off time) each having three levels. Remaining machining parameters such as polarity, flushing pressure, and voltage are set at fixed setting for whole experimental work. Total nine runs with three replications are performed during experimentation.

The result of responses converted into signal-to-noise (S/N) ratio. Signal-to-noise (S/N) ratio is calculated based on "higher is better" for MRR. Higher is better deals with more material which is required to be removed from workpiece while machining. Table 3 shows the experimental design, into which S/N ratio of MRR is calculated at different parameters combination according to L_9 orthogonal array design scheme.

4 Results and Discussion

Analysis of variance (ANOVA) is applied on output results to find out the significance level and contribution of each input parameter. For optimization, optimum level of each parameter is calculated using the analysis of means (ANOM). Table 4 shows the results of ANOVA.

Current is found the most dominating factor for MRR with 97.8% contribution. However, pulse on time and pulse off time both parameters have negligible contribution. Results of ANOM are given in Table 5 which shows the optimal settings for maximum MRR. Higher MRR can be achieved by setting the current and pulse on time (B) at level 3 but pulse off time (C) at level 1.

The influence of machining parameters on response material removal rate which is calculated based on higher is better S/N ratio. It is observed that MRR increased with increasing current and pulse on time as shown in Fig. 2 and Fig. 3, respectively.

Table 4 Results of ANOVA and contribution of each parameters

Parameters	Degree of freedom	Sum of squares	Mean sum of squares	F-ratio	Contribution % age
Current	2	1336.33	668.16	2770.32	97.89
Pulse on time	2	21.20	10.60	43.96	1.518
Pulse off time	2	2.17	1.08	4.50	0.12
Error	20	4.82	0.24		0.45
Total	26	1364.54			100

Table 5 Results of ANOM for MRR

Level	A		B		C	
	Raw data	S/N Ratio	Raw data	S/N Ratio	Raw data	S/N Ratio
1	1.38	2.65	9.05	14.69	*10.48*	15.81
2	10.29	20.16	10.01	16.24	9.99	15.96
3	*18.60*	*25.38*	*11.21*	*17.27*	9.80	*16.42*

Fig. 2 Current versus MRR

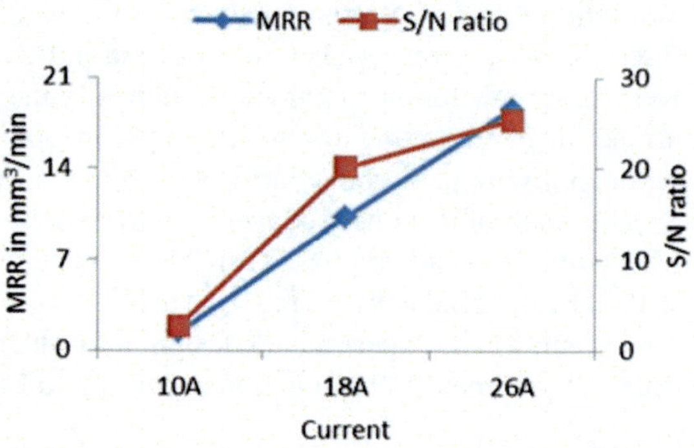

Fig. 3 Pulse on time versus MRR

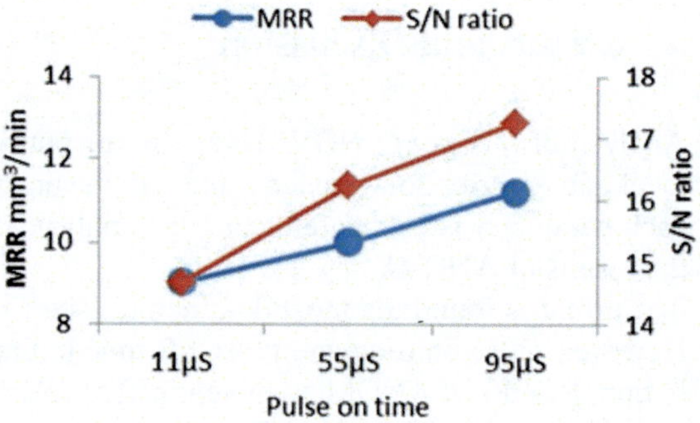

Fig. 4 Pulse off time versus MRR

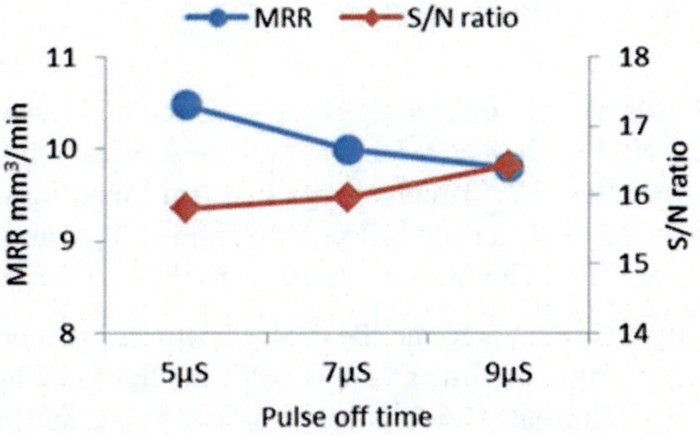

Current provides energy in the form of spark during machining, so higher energy will remove more material from the working zone. However, high pulse on time gives sufficient time to generate the spark. Low pulse off time gives more MRR value which is shown in Fig. 4.

5 Confirmation Run

The confirmation run is necessary to validate the predicted and experimental results at optimum levels. The predicted value of MRR at optimal level of each input parameter is measured by using following expression:

$$\mu = \Psi m + \sum_{k=1}^{n} (\Psi m_i - \Psi m) \tag{3}$$

where μ is the estimated mean value of response, Ψm is the overall mean of response, Ψm_i is the mean of optimum level of ith significant factor, and n is the number of significant parameters. The results of confirmation run are given in Table 6. The estimated responses are calculated from S/N ratio for MRR. Confidence interval (C.I.) at 95% confirms the experimental responses. The range of C.I. is used to validate the experimental results value for best parameters.

Table 6 Results of confirmation run at 95% confidence interval

Confirmation test for MRR				
	Parameters with level	Predicted	Experimented	Confidence interval
Optimum settings	$A_3B_3C_1$s	20.71	18.85	18.73–22.9
Worst settings	$A_1B_1C_1$	–	1.07	–
Improvement in MRR (mm^3/min)			17.78	

6 Conclusions

The effect of machining parameters on MRR is evaluated in this experimental work. ANOVA and ANOM techniques are applied to find out the effect of input parameters on MRR. Experimental data is transformed into signal-to-noise (S/N) ratio taking higher is better for MRR. Validation of optimization is done through confirmation run. The following conclusions are drawn from experimental work:

1. Current is found the most significant parameter to improve MRR.
2. Pulse on time and pulse off time have negligible effect on MRR.
3. Current at 26 A, pulse on time at 95 μs, and pulse off time at 5 μs are the optimal level for MRR.
4. MRR is improved with the amount of 17.78 mm^3/min at optimum settings.
5. Improvement of MRR reveals the success of Taguchi approach to optimize the process.

References

1. Abbas NM, Solomon DG, Bahari MF (2007) A review on current research trends in electrical discharge machining (EDM). Int J Mach Tools Manuf 47(7–8):1214–1228. https://doi.org/10.1016/j.ijmachtools.2006.08.026
2. SinghAK KS, Singh VP (2014) Optimization of parameters using conductive powder in dielectric for EDM of super Co 605 with multiple quality characteristics. Mater Manuf Processes 29(3):267–273. https://doi.org/10.1080/10426914.2013.864397
3. Lin YC, Cheng CH, Su BL, Hwang LR (2006) Machining characteristics and optimization of machining parameters of SKH 57 high-speed steel using electrical-discharge machining based on Taguchi method. Mater Manuf Processes 21(8):922–929. https://doi.org/10.1080/03602550600728133
4. Ho KH, Newman ST (2003) State of the art electrical discharge machining (EDM). Int J Mach Tools Manuf 43(13):1287–1300. https://doi.org/10.1016/S0890-6955(03)00162-7
5. Jain VK (2002) Advanced machining processes, Allied Publishers, New Delhi, ISBN 81-7764-294-4
6. Chattopadhyay KD, Verma S,Satsangi, PS, Sharma PC (2009) Development of empirical model for different process parameters during rotary electrical discharge machining of copper–steel (EN-8) system. J Mater Process Technol 209(3):1454–1465. https://doi.org/10.1016/j.jmatprotec.2008.03.068
7. Singh AK, Kumar S, Singh VP (2015) Effect of the addition of conductive powder in dielectric on the surface properties of superalloy Super Co 605 by EDM process. Int J Adv Manuf Technol 77(1):99–106. https://doi.org/10.1007/s00170-014-6433-z
8. Chattopadhyay KD, Satsangi PS, Verma S, Sharma PC (2008) Analysis of rotary electrical discharge machining characteristics in reversal magnetic field for copper-en8 steel system. Int J Adv Manuf Technol 38(9):925–937. https://doi.org/10.1007/s00170-007-1149-y
9. Jeyapaul R, Shahabudeen P, Krishnaiah K (2005) Quality management research by considering multi-response problems in the Taguchi method—a review. Int J Adv Manuf Technol 26(11):1331–1337. https://doi.org/10.1007/s00170-004-2102-y
10. He Z, Zhu PF, Park SH (2012) A robust desirability function method for multi-response surface optimization considering model uncertainty. Eur J Oper Res 221(1):241–247. https://doi.org/10.1016/j.ejor.2012.03.009

11. Parashar V, Purohit R (2017) Investigation of the effects of the machining parameters on material removal rate using Taguchi method in end milling of steel grade EN19. Mater Today Proc 4(2):336–341. https://doi.org/10.1016/j.matpr.2017.01.030
12. Singh K, Singh AK, Chattopadhyay KD (2020) Application of Taguchi method to optimize the surface roughness during face milling of rolled steel (AISI 1040). In: Proceedings of international conference in mechanical and energy technology, pp 239–249. Springer, Singapore. https://doi.org/10.1007/978-981-15-2647-3_2

Parametric Influence on Material Removal Rate of K-90 Alumina Ceramics During Zircon Sand (Abrasives) Jet Assisted Machining

Santosh Kumar Sahu and Swastik Pradhan

Abstract Abrasive jet machining (AJM) is a non-traditional machining process where removal of material from the workpiece takes place due to application of high-speed stream of abrasive particles carried by high-pressure air. Material removal takes place due to impact erosion of hard abrasive particles which cause brittle fracture of targeted tiny surface of the workpiece. In this paper, an experimental analysis is carried out to investigate material removal efficiency of AJM on K-90 Alumina ceramic in consideration with Zircon sand as abrasives. Effects of AJM parameters including standoff distance (4, 6, 8 mm), air pressure (3, 5, 7 kgf cm^{-2}), abrasive grain size (average size 180, 500, 710 μm), and abrasive temperature (60, 80, 100 °C) on material removal rate are studied. Response surface methodology (RSM) is applied to derive a second-order mathematical relationship representing material removal rate as a function of selected AJM parameters. Beneficial effect in utilizing hot abrasives is evidenced in purview of improved material removal efficiency.

Keywords AJM · K-90 alumina ceramic · Zircon sand · MRR · RSM

1 Introduction

AJM is one of the non-conventional machining routes where variety of operations like burr removal, drilling, surface polishing, etc., can be carried out in an efficient manner with very close tolerance [1]. Venkatesh studied AJM performance on glass, ceramics, and die steel workpieces in consideration with process variants: nozzle tip distance (NTD), spray angle, and pressure. SiC and Al_2O_3 were used as abrasives; nozzle was made of tungsten carbide. Elliptical and bell-mouthed drilled holes were formed instead of circular and cylindrical [2]. Wakuda et al. reported that AJMed

S. K. Sahu (✉)
Department of Mechanical Engineering, Veer Surendra Sai University of Technology, Burla, Odisha 768018, India

S. Pradhan
School of Mechanical Engineering, Lovely Professional University, Punjab 144411, India

A. K. Dubey et al. (eds.), *Recent Trends in Industrial and Production Engineering*, Lecture Notes in Mechanical Engineering,
https://doi.org/10.1007/978-981-16-3135-1_19

surface did not experience strength degradation as cracks that formed radially were not oriented downwards because of particle impact upon ceramic work material. Authors recommended AJM to execute damage-free micro-machining of ceramic materials [3].

Park et al. studied effect of micro-grooving upon glass during micro-AJM with the help of masking process (laminating, exposure, developing, AJM process, mask removing, and cleaning), and film thickening techniques by considering process parameters: standoff distance (SOD), blasting air pressure, inner nozzle diameter, and flow rate of micro-abrasives (WA) alumina [4]. Mahajan studied influence of AJM parameters (abrasive size, standoff distance, and pressure) on MRR, and drilled diameter on glass plates by using SiC abrasives. It was experienced that increase in SOD caused wider top, and bottom diameter of the hole produced. In addition, increase in pressure improved material removal efficiency [5]. Nanda et al. performed abrasive jet drilling of borosilicate glass using different grades of zircon sand (abrasives) in consideration with varied air pressure. Effects of process parameters: pressure, standoff distance, grain size, and temperature on MRR, surface roughness, and flaring diameter were analyzed through response surface methodology [6]. Nassef et al. studied parametric influence (nozzle diameter, standoff distance, grain size, and pressure) on *kerf* during AJM drilling of glass. It was witnessed that increase in SOD caused increased kerf taper which adversely affected dimensional accuracy. Literature is rich enough focusing aspects of material removal rate of AJM in consideration with variety of abrasives [7]. To cope up with present-day industrial needs to manufacture products having complicated part geometry with tight dimensional tolerance, superior surface finish, especially, from hard and brittle materials, AJM is being widely attempted. Very few researches were carried upon the effect of hot air abrasives on the workpiece [8]. Majority of past studies utilized abrasives at room/laboratory temperature. In the present experimental, an attempt is made to study effects of application of hot abrasives (carried by compressed air) on MRR of alumina ceramic.

2 Experimental Procedure

The experimental setup is described in Fig. 1. In this work, K-90 alumina ceramic (dimension $25 \times 25 \times 4$ mm^3 is taken as workpiece. Zircon sand (ZrO_2) is used as abrasives with varied mesh size no (corresponding to grain size 180, 500, and 710 μm, respectively). Table 1 exhibits composition of K-90 alumina ceramic.

ZrO_2 powder is chosen as abrasives due to its hardness. It does not crumble into parts while striking the work surface; thus, there is less chance of abrasive sticking onto the work surface. It has high abrasion resistance and is relatively cheap. It is used in various purposes like slot cutting, surface finishing, surface cleaning, and drilling of hard workpiece. Its color is brownie red (Fig. 2). Properties of ZrO_2 are furnished in Table 2. Constituents of zircon sand is provided in Table 3.

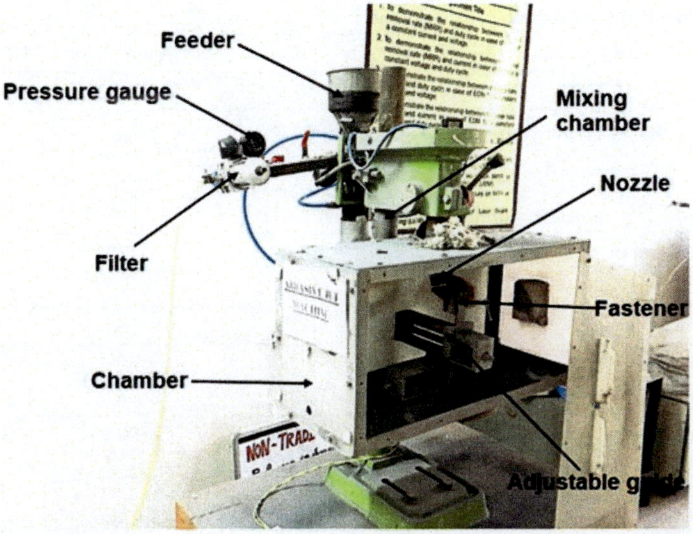

Fig. 1 Experimental setup

Table 1 Composition of K-90 alumina ceramic material

Al_2O_3	SiO_2	CaO	Na_2O	Others
≥92.00%	≥5.62%	≤1.40%	≤0.1%	≥0.8%

Source http://www.xieta.com

Fig. 2 Zircon sand (180, 500, and 710 μm)

Table 2 Constituents of zircon sand

Physical properties of zircon sand	
Bulk density (Uncompacted)	2643–2804 kg/m^3
Specific gravity	4.60–4.71
Hardness (MOHS)	7.5
Melting point	2100–2300 °C
Coefficient of linear expansion	7.2 × 10^{-6} cm/cm °C

Table 3 Chemical composition of zircon sand

ZrO_2	SiO_2 (Total)	Fe_2O_3	Na_2O	CaO	Hf
62.90%	31.23%	3.77%	0.30%	0.08%	1.50%

Source https://www.911metallurgist.com

Table 4 Process parameters with levels of variation

Parameters/Symbol	Unit	Levels		
		-1	0	+ 1
Air pressure, A	$kgf.cm^{-2}$	3	5	7
Temperature, B	°C	60	80	100
Stand-off distance, C	mm	4	6	8
Abrasive grain size, D	μm	180	500	710

For conducting experiments, fluidized bed (FB) hot abrasive jet machining arrangement is prepared; abrasive chamber (to ensure air-abrasive mixing), abrasive heating arrangements with thermocouple, stainless steel nozzle having high abrasion (wear) resistance, etc., are equipped with the AJM setup. In this experiment, nozzle diameter (6 mm) is kept fixed. The following process variables (Table 4) are considered: air pressure, standoff distance, abrasive size, and temperature. Mass loss of the workpiece is measured using electronic balance (ER-180A, AFCOSET Co. Ltd., India). It has capacity to weight up to 180 g with accuracy reading 0.001 mg. Process performance is assessed in terms of material removal rate (MRR). Levels of parametric variation and corresponding values are selected after performing few pilot tests by adjusting standoff distance along with the amount of work material removed. Design of experiment is based on Box–Behnken design (Table 5). Experimental layout* and MRR data are provided in Table 5.

The volume of removed unwanted stuff from the workpiece per unit time is defined as MRR. Due to impact erosion of workpiece by abrasive particles having high kinetic energy, material removal takes place from the workpiece. Hence, MRR can be computed as difference between initial and final mass of the workpiece, divided by machining duration. Machining duration is taken as 20 s for all experimental trials. The snapshot of machined workpiece is shown in Fig. 3.

$$\text{MRR} = \left(W_{in} - W_{fi}\right)/t.$$

In which, W_{in} = initial mass of the workpiece.
W_{fi} = final mass of the workpiece after machining.
t = machining duration (s).

Table 5 Design of experiment and collected data

Run number	DOE, Box-Behnken design				MRR × 10^3 g/s
	A	B	C	D	
1	0	0	−1	−1	2.11
2	0	0	1	−1	6.05
3	0	1	1	0	7.65
4	1	-1	0	0	3.99
5	1	1	0	0	6.25
6	−1	0	1	0	2.99
7	−1	−1	0	0	2.32
8	1	0	0	−1	4.51
9	0	0	0	0	2.81
10	0	0	1	1	5.99
11	0	0	−1	1	3.32
12	1	0	1	0	6.05
13	1	0	0	1	4.99
14	−1	0	−1	0	2.05
15	0	−1	0	1	4.01
16	0	−1	−1	0	2.04
17	0	1	0	1	5.51
18	0	1	−1	0	2.51
19	0	1	0	−1	3.81
20	−1	0	0	−1	2.49
21	0	−1	0	−1	3.99
22	0	0	0	0	4.00
23	0	−1	1	0	4.79
24	−1	1	0	0	2.75
25	1	0	−1	0	3.39
26	−1	0	0	1	2.51
27	0	0	0	0	3.48

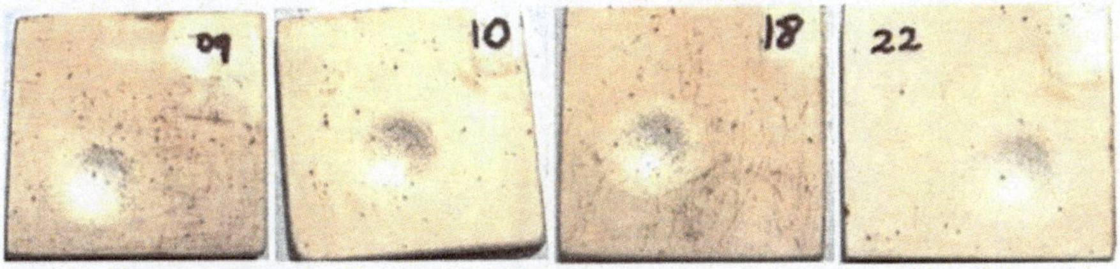

Fig. 3 Snapshot of machined workpieces

3 Result and Discussion

Response outputs, i.e., MRR (Fig. 4), obtained according to Box–Behnken design of experiment, are analyzed through MINITAB 19 software. Through RSM technique, a quadratic mathematical model is developed in order to find out the best fit between response variable and given input parameters with coefficient of determinations (experimental and adjusted). In order to estimate parametric significance (direct, and interactive effect) on response variable (MRR, in the present case), analysis of variance (ANOVA) is applied. From Table 6, values of P for those variables which are less than 0.05 in the fitted model indicates error percentage ~5% (or less than that). This means presence of those variables in the model which is acceptable at 95% confidence level.

From Fig. 5, surface plot between MRR and temperature (T), it can be observed that with increase in abrasive temperature level the value of MRR increases accordingly; this is due to the fact that refinement of grains and sharpening its outer edge causes more materials to remove upon impact with the workpiece. MRR is very high at peak level of temperature as compared to the lower temperature level because machined surfaced can be the plastic deformation due to hot abrasive grain. In between MRR and pressure, it can be observed that that with the increase in the pressure the MRR value increases accordingly this is due to the fact that high pressure causes high kinetic energy and high mass flow rate of abrasive particle impacts upon the workpiece causing a high MRR during the time interval. In between MRR and grain size, it can be observed that with the increase in the grain size MRR increases; the reason behind this is that bigger size grain causes greater impact upon unit surface area as compared to smaller size grain, so the MRR is higher in case of larger grain as compared to smaller grain. Again, from Fig. 5, it can be observed that with the increase in the standoff distance, MRR value increases, reaches up to a level, and then slightly drops; the reason behind this is that as the standoff distance

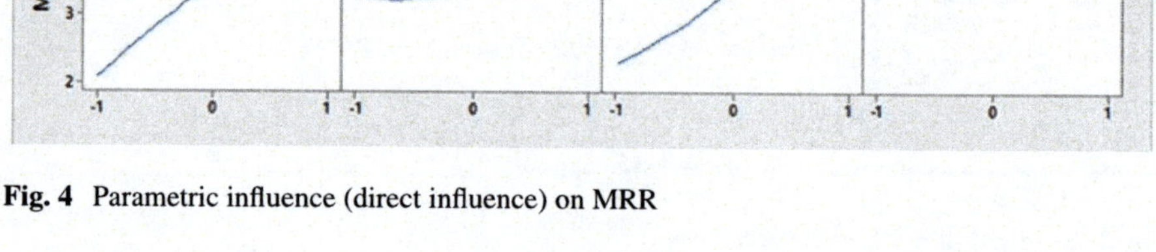

Fig. 4 Parametric influence (direct influence) on MRR

Table 6 ANOVA: Response surface regression: MRR versus *A*, *B*, *C*, *D*

Term	Effect	Coef	SE Coef	*T*-Value	*P*-Value	VIF
Constant		3.430	0.335	10.22	0.000	1.00
A	2.345	1.173	0.168	6.99	0.000	1.00
B	1.223	0.612	0.168	3.65	0.003	1.00
C	3.017	1.508	0.168	8.99	0.000	1.00
D	0.562	0.281	0.168	1.67	0.120	1.25
A * *A*	−363	−0.182	0.252	−0.72	0.484	1.25
B * *B*	0.969	0.485	0.252	1.93	0.078	1.25
C * *C*	0.799	0.400	0.252	1.59	0.138	1.25
D * *D*	0.887	0.443	0.252	1.76	0.103	1.00
A * *B*	0.915	0.457	0.291	1.57	0.141	1.00
A * *C*	0.860	0.430	0.291	1.48	0.165	1.00
A * *D*	0.230	0.115	0.291	0.40	0.699	1.00
B * *C*	1.195	0.597	0.291	2.06	0.062	1.00
B * *D*	0.840	0.420	0.291	1.45	0.174	1.00
C * *D*	−0.635	−0.317	0.291	−1.09	0.296	1.00

$\text{MRR} = 3.430 + 1.173\,A + 0.612\,B + 1.508\,C + 0.281\,D - 0.182\,A*A + 0.485\,B*B + 0.400\,C*C + 0.443\,D*D + 0.457\,A*B + 0.430\,A*C + 0.115\,A*D + 0.597\,B*C + 0.420\,B*D - 0.317\,C*D$

increases the MRR increases due to high velocity of abrasives and kinetic energy but further increasing in standoff distance causes the drag force of atmosphere that comes into play which reduces the velocity of flow of abrasive for which MRR decreases.

4 Conclusion

- From the ANOVA table of MRR, it was accomplished to find out the magnitude of the quadratic models by considering the '*P*' value less than 0.05 and the preconceived *F* value greater than the tabular value of *F*. Here, from the ANOVA table, it was found that the regression model and temperature are found to be the significant models. It was also found from the surface plot that increase in process parameters increases MRR.
- The anticipated optimal amalgamation of parameter setting is pressure of 5 Kgf/cm^2, standoff distance of 8 mm, grain size of 500 μm, and temperature of 100 °C for achieving the optimal output, i.e., MRR of 0.00765 gm/s (Run No-3).
- From the surface plots for MRR, we found that MRR value increases with increase in the input process parameters like MRR increases with increase in pressure, temperature, and grain size, and for SOD, it increases and further decreases slightly.

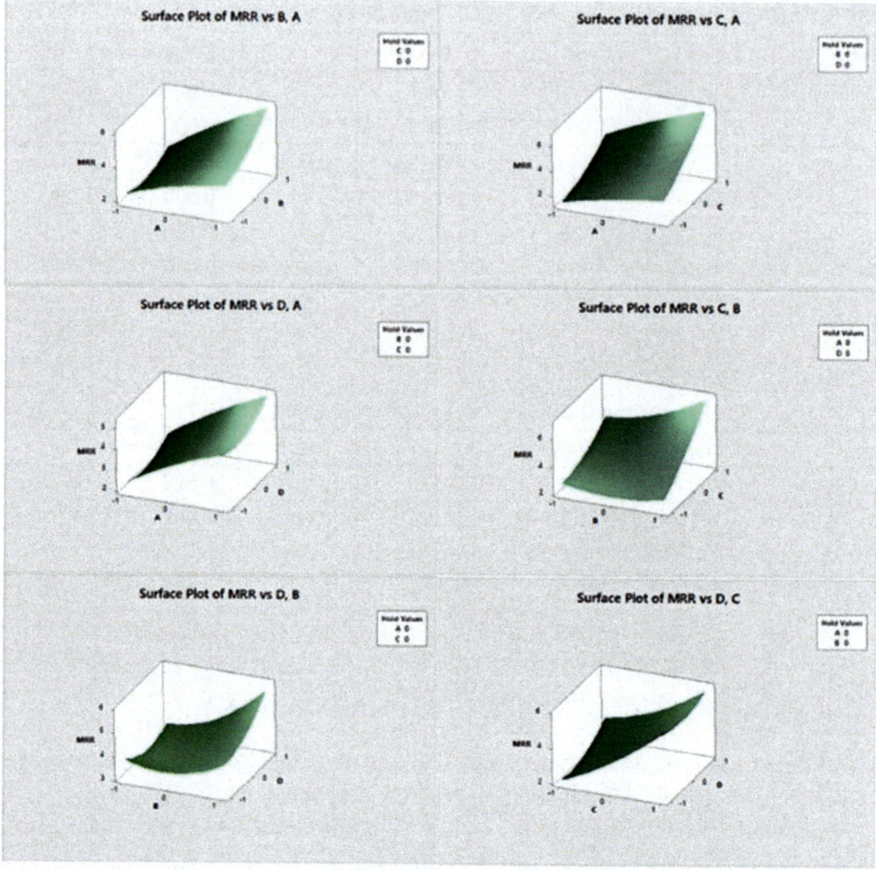

Fig. 5 Interactive effects of parameters on MRR

- The value of *R*-square from regression analysis is 93.32% which indicates that the model is best fitted for the input process parameters.

References

1. Ramachandran N, Ramakrishnan N (1993) A review of abrasive jet machining. J Mater Process Technol 39(1–2):21–31. https://doi.org/10.9790/1684-12424452
2. Venkatesh VC (1984) Parametric studies on abrasive jet machining. CIRP Ann 33(1):109–112. https://doi.org/10.1016/S0007-8506(07)61390-0
3. Wakuda M, Yamauchi Y, Kanzaki S (2002) Effect of workpiece properties on machinability in abrasive jet machining of ceramic materials. Precis Eng 26(2):193–198. https://doi.org/10.1016/S0141-6359(01)00114-3

4. Park DS, Cho MW, Lee H, Cho WS (2004) Micro-grooving of glass using micro-abrasive jet machining. J Mater Process Technol 146(2):234–240. https://doi.org/10.1016/j.jmatprotec.2003.11.013
5. Mahajan G (2014) A study of effect of various process parameters on abrasive jet machining using silicon carbide as abrasive material. Int J Eng Dev Res 3:25–31. http://www.ijedr.org/papers/IJEDR1501006.pdf
6. Nassef A, Elkaseer A, Abdelnasser ES, Negm M, Qudeiri JA (2018) Abrasive jet drilling of glass sheets: effect and optimization of process parameters on kerf taper. Adv Mech Eng 10(1):16–35. https://doi.org/10.1177/1687814017748435
7. Nanda BK, Mishra A, Dhupal D (2017) Fluidized bed abrasive jet machining (FB-AJM) of K-99 alumina ceramic using SiC abrasives. Int J Adv Manuf Technol 90(9):3655–3672. https://doi.org/10.1007/s00170-016-9699-5
8. Pradhan S, Das SR, Dhupal D (2020) Performance evaluation of recently developed new process HAJM during machining hardstone quartz using hot silicon carbide abrasives: an experimental investigation and sustainability assessment. Silicon, Springer (in Press). https://doi.org/10.1007/s12633-020-00641-9

Recent Development in Friction Stir Welding: An Advancement in Welding Technology

Akash Sharma and Vijay Kumar Dwivedi

Abstract Friction stir welding (FSW) has various applications focused on the different industries because of its efficiency and sustainability. The main different sectors of industries where FSW technique is utilized are automobiles, aerospace, shipbuilding and railways. This current experimental study describes about the background of the FSW technique and recent advances in welding technology which have occurred during the last few years. Furthermost, the development endeavored to conflict the problem of non-uniformity in temperature distribution in weld zone pinhole, which is left after the welding process in FSW. During the last two decades, many researchers have been working on the advanced welding technology of FSW because this technique could be made more optimized and sophisticated by using proper tool during the welding process. The application of various parameters such as tool rotation, plunge depth, feed rate, shoulder diameter of tool and tool pin geometry has made FSW process more uniform. Friction stir welding process is a cutting edge technology widely used nowadays throughout the globe for welding because of its very high strength. The joint formation in the process has a wide application in welding of two dissimilar materials, e.g., aluminum and steel. In this research paper, an attempt is made to provide some directions for future development in the friction stir welding technology.

Keywords Friction stir welding · Temperature distribution · Defects in FSW

1 Introduction

Friction stir welding is a strong state thermo-mechanical solid state joining of material which can be similar or dissimilar (a combination of expelling and fashioning), designed by The Welding Institute (TWI) in 1991, that has become a feasible assembling innovation of metallic sheet and plate materials for applications in different ventures, including plate materials for applications in different businesses, including

A. Sharma (✉) · V. K. Dwivedi
Mechanical Engineering Department, GLA University, Mathura 281406, India

A. K. Dubey et al. (eds.), *Recent Trends in Industrial and Production Engineering*, Lecture Notes in Mechanical Engineering,
https://doi.org/10.1007/978-981-16-3135-1_20

aviation, vehicle, safeguard and shipbuilding. Friction stir welding (FSW) is a generally new procedure created by The Welding Institute (TWI) for the joining of Aluminum combinations [1]. Friction stir welding process is nowadays widely used throughout the globe as an advanced welding technology which produces high strength weld joints.

Solid state joining process (FSW) measure is generally another advanced joining process that is by and by pulling in extensive intrigue. FSW process is nowadays trending and widely accepted throughout the manufacturing industries for high strength welding process. FSW is rising as a suitable elective technology with high effectiveness because of high-handling speeds. Since the joint can be attained underneath the dissolving temperature, this technique is reasonable for joining various materials those are amazingly hard to be welded by conventional welding techniques used commonly in the industries [2, 3]. The welding is having appealing state in nature and depends on the limited fashioning of the desired weld zone to deliver constituently high strength and defect-free joint as appeared in Fig. 1.

Friction stir welding produces welds by utilizing a revolving, non-consumable welding instrument to locally make softer a workpiece through which the material of both the workpiece get mixed with each other in the weld zone, through the high temperature heat created by contact and plastic work, accordingly permitting the tool to "stir" the joint surfaces. In friction stir welding cycle, a rotating welding tool is rotated with very high and controlled revolutions penetrates the material at the edge of specimen material, for instance, two abutting plates of same or dissimilar material, and afterward deciphered along the interface. FSW a solid state welding process offers simplicity of taking care of, exact outside cycle control and elevated levels of repeatability, accordingly making extremely homogenous welds. No exceptional planning of the example is required and very minimal waste or contamination is made during the process of welding cycle of material. Moreover, its relevance to

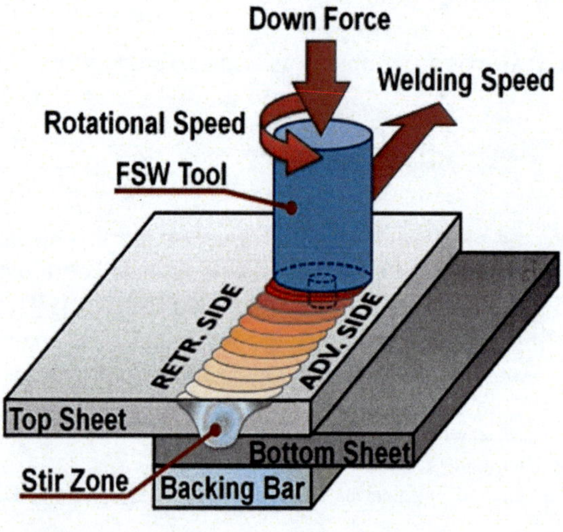

Fig. 1 Schematic sculpt of FSW process

aluminum combinations, in standard particular unique joints or those considered "unweldable" by traditional welding procedures, for example, tungsten inert gas (TIG) welding, make it as an appealing strategy for the transportation division. The friction stir welding process measure includes the interpretation of a high RPM rotating cylindrical shaped tool along the interface between two plates which is to be joined by using the non-conventional friction stir welding process. Friction due to the rotating tool uniformly produces a high temperature which results in heating the material to its softening stage at which the material can be mixed with each other, which is then basically expelled around the tool before being fashioned by the enormous down weight [4]. The weld produced due to the high temperature generated by friction is shaped by the deformation of the specimen material at temperature underneath the softening temperature. The synchronous rotational and translational movement of the welding instrument during the friction stir welding cycle makes a qualities asymmetry between the bordering sides. On one hand, where the tool turn is with the bearing of the interpretation of the welding instrument one pinnacles of the advance side (AS), while then again, the two different movements, pivot and interpretation neutralize and one talks about the retreating side (RS) [5, 6]. Friction stir welding is an advanced process of joining of materials with high strength, Calibrated sensors used for measuring the temperature (thermocouple) in the process make the process more uniform and accurate with very less chances of defects which generally takes place due to the fluctuation in the weld zone temperature. In friction stir welding process when the tool plunges in the material, it requires a uniform temperature to carry out the whole welding process defect less and very high strength weld joint is produced.

Lately, mathematical displaying of FSW has given noteworthy understanding about the warmth age designs, materials stream fields, temperature profiles, lingering pressure and contortion, and certain parts of hardware plan [7]. The advancement of proper tool materials for welding process and calculations has made friction stir welding conceivable to join the materials, for example, steel and titanium in the research center condition and in a set number of creation applications [8]. In friction stir welding process, of steel material, it has been demonstrated that the lower welding temperature can prompt low bending and interesting joint properties. FSW of steel is a region of dynamic exploration, that is why the process is sensible to anticipate that other creation applications should rise after some time. An extremely appealing application of FSW process can be well known as welding of steel plate for shipbuilding applications, in light of on the decrease of welding contortion, and however, the improvement of minimal effort welding hardware and more vigorous welding instrument materials is required before this application of welding process can be started.

The Accucut machine which is having very good capability of contributing in the friction stir welding process because it has several parameters and arrangements which helps in executing the friction stir welding process much smoother. Accucut machine is used frequently by using different fixtures as per the requirement of friction stir welding process. The processing machine utilized during this trial is Accucut as appeared in Fig. 2 [9].

Fig. 2 Accucut milling machine used during FSW process

2 Material Used

AA 7075-T6 aluminum alloy material plates having a thickness of 6 mm is chosen for the experimental work which is carried out in this research paper. Workpiece is having specific size which comprises of 150 mm × 125 mm × 6 mm for the experimentation work in friction stir welding measure. This test has been acted in the division of mechanical designing (University Polytechnic), GLA University, Mathura, India on an Accucut vertical processing machine to perform FSW. The substance organization by weight (wt%) of AA-7075 utilized for playing out the rubbing stir welding (FSW) test is appeared in Table 1. A non-consumable (FSW) tool with an alternate shapes and size made of En-31 is utilized to play out the weld

Table 1 Chemical composition of AA7075 according to spectrometer analysis (wt%)

Element	Si	Fe	Cu	Mn	Mg	Al	Cr	Zn	Ti
Required	0.3	0.4	1.2–	0.3	2.0–	87.1–	0.19–	5.0–	0.3
wt%			2.0		2.8	90.2	0.29	6.2	

Table 2 Mechanical property of AA7075-T6 according to spectrometer analysis

Brinell hardness	150
Ultimate tensile strength (MPa)	572
Yield strength (MPa)	503
% Elongation	11
Modulus of elasticity (MPa)	70.7
Fracture toughness (Joule)	20
Poisson ratio	0.33
Mach inability	70%

joints at different speed by erosion stir welding. The dimensions of the tool used in this project are as follow, tool shoulder width of 20 mm, pin measurement of 5 mm at the root and pin length 5.75 mm is utilized for FSW measure. In FSW, rotating tool heat creates to get bringing about nearby plastic disfigurement. The heat is produced during this process of friction stir welding process due to the friction generated due to the rotation of the tool whose tool pin is penetrated inside the plate of specific material which is to be joined [9] (Table 2).

3 Temperature Measurement During FSW

The plates were set up to quantify the temperature at four focuses utilizing thermocouples enwrapped by fired cylinder. On each plate, two K-type thermocouples of 0.5 mm distance across were welded to the surface. The areas of the thermocouples on the workpiece are appeared underneath in Fig. 3. Location A, B, C, D is having thermocouples once the temperature sensors (thermocouple) are not calibrated and

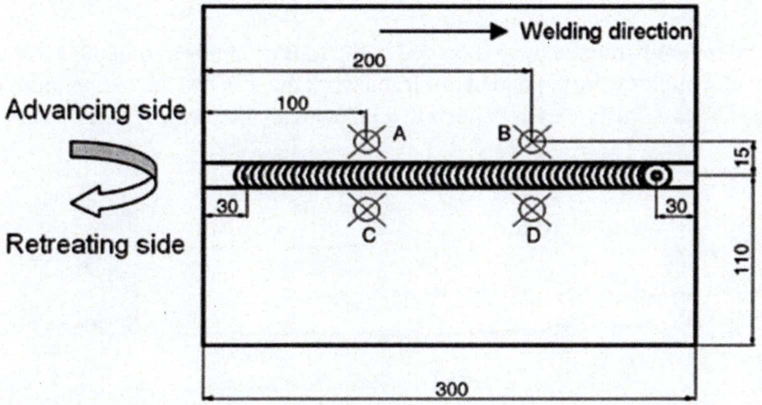

Fig. 3 Location of thermocouples for temperature measurement during FSW process

again with that same process parameters when the temperature sensors are calibrated. The calibrated temperature sensors are very effective in obtaining the accurate and precise temperature.

3.1 *Thermocouple Used for Measurement*

Type K (chromel–alumel) is the most well-known broadly useful thermocouple with an affectability of roughly 41 μV/°C as appeared in Fig. 4. Type K thermocouple is economical, and a wide range of experiments of tests related to temperature profile are accessible in its −200 to +1350 °C (−330 to +2460 °F) extend. Sensitivity of Type K thermocouple is good for measurement within this range of temperature (−200 to 1200) °C. Type K thermocouple is used during this experimental study because it can be easily made by taking two different wires of chromel and alumel. These wires used in thermocouples are easily available in the market. Type K thermocouple can be easily used by affixing the temperature sensor (thermocouple) on the different points of different weld zones from which the temperature profile is obtained and accordingly the welding process is made smoother with avoiding the various defects occurred due to the uncontrolled temperatures.

Type K thermocouple is having very high sensitivity which results in sensing the temperature within very less time period. The thermocouple is very easy to affix at the different zones where the temperature profile is required because the wires are having such strength that it cannot be broken easily by very less applied force. Thermocouple which is calibrated results in making a very accurate and precise temperature profile which will contribute in making the friction stir welding process defect-free and the weld zone after the process is completed is having high strength and good surface finish.

The thermocouples can't be put in the thermo-precisely influenced zone (TMAZ) of the weld since the blending of the tool will crush it when the tool arrives at its area.

Transient temperatures were recorded in the four channels during the FSW cycle.

Thermocouples were appended to a framework that can test the temperature information. The last thusly was appended to a PC with an altered program for recording

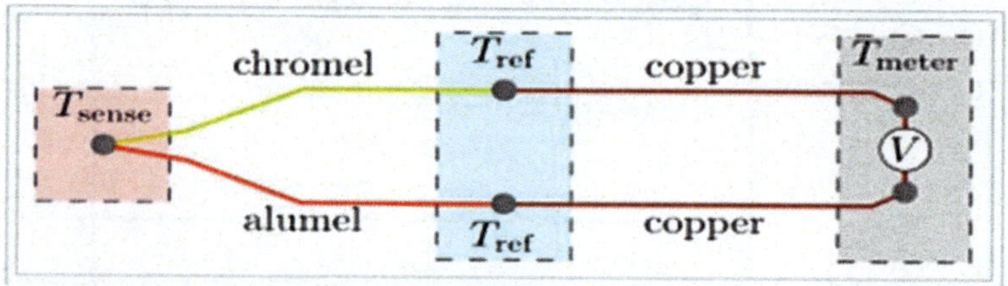

Fig. 4 Internal design of Type K thermocouple

the temperature information. The Type K thermocouple is first calibrated and then it is placed on the different location of workpiece for sensing the temperature. The indicator is also calibrated along with the TC for Accurate measurement of temperature during the Heat generation in the FSW Process. Calibration plays a vital role through which an accurate measurement can be done and a very precise temperature profile of FSW weld zone temperature can be drawn, so that the different defects caused through overheating can be minimized as shown in Fig. 5. The temperature profile helps in maintaining the required temperature throughout the process which makes the FSW welding process more smooth and reliable.

Maximum temperatures recorded for same FSW workpiece without calibration and after Calibration with same process parameters of FSW process as shown in Table 3 [10, 11]. In Table 3, it is seen that there is a wide difference in the temperature based on the sensor, calibrated sensor is giving exact temperature and the other sensor without calibration is showing some other temperature value due to which, the welding process is suffered and the welding process is not that much effective. Temperature profile which is obtained through calibrated temperature sensor is considered for better and precise friction stir welding process.

The step-shaped workpieces. Uncontrolled welding due to the varying temperature beyond the suitable temperature range contributes in a meltdown of the tool due to the uncontrolled temperature beyond the limit, with the temperature controller enabled, the narrow zone of the Al alloy weld specimen is successfully passed as shown in

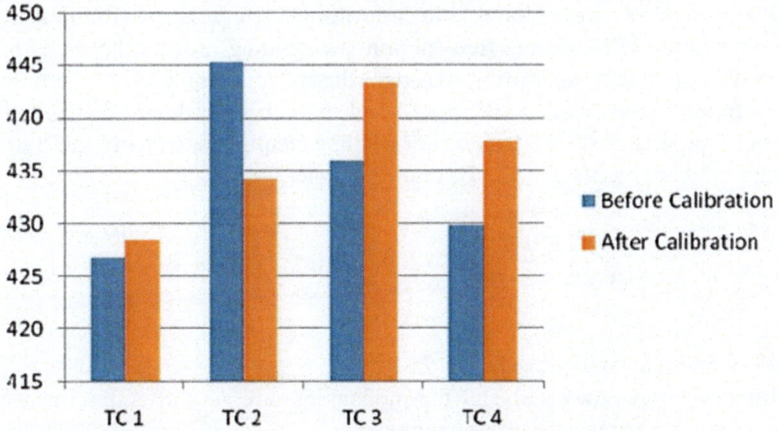

Fig. 5 Temperature profile before and after calibration of TC (Thermocouple)

Table 3 Temperature profile before and after calibration of TC (Thermocouple)

	T1 °C	T2 °C	T3 °C	T4 °C
Before calibration	426.8	445.3	435.9	429.9
After calibration	428.5	434.2	443.4	437.8

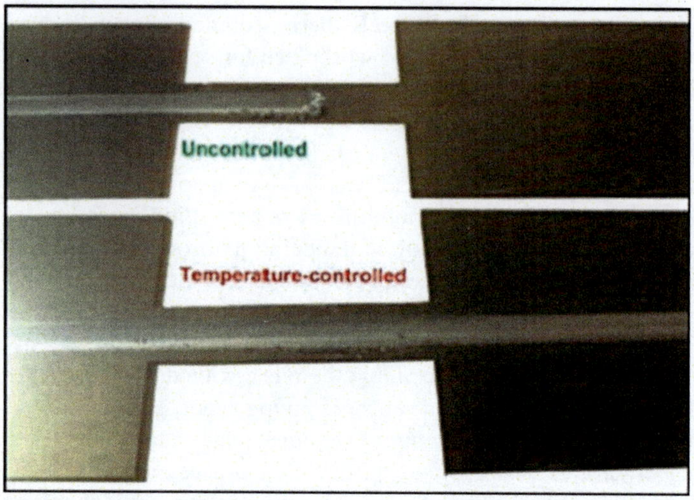

Fig. 6 FSW process with controlled and uncontrolled Temperature

Fig. 6. The temperature during the whole process of friction stir welding should be constantly monitored as well as the temperature should not cross the boundary line, which will affect the weld zone and the various defects will occur in the welded workpiece.

Calibration plays a very vital role in controlling a temperature within the desired suitable temperature for defect-free friction stir welding process, because through calibrated temperature measuring sensors (thermocouples) accurate temperature of the different zones can be monitored and as well accordingly controlled. The controlled temperature during friction stir welding results in defect-free, high strength welding with better surface finish of the weld zone.

4 Conclusion

- FSW opening up new areas of welding daily.
- Calibrated temperature sensors (thermocouples) are very effective for accurate temperature monitoring and controlling.
- In this research, the distribution of temperature throughout the workpiece Al 7075 T-6 is analyzed experimentally during the FSW process.
- Accurate temperature measurement methods applied to the FSW process reduces various defects caused by overheating in weld zone like blur spots, voids, etc.
- It is determined from the above experiment, the temperature on the Advancing side (AS) of the weld seems to be bit high comparatively to the Retreating side (RS) of the weld.

- Though it is incredibly hard to quantify the temperatures at the weld line, an endeavor has been made to decide the temperatures around the edge of the tool shoulder.
- From the examination, it very well may be presumed that the proper temperature for a deformity free friction stir weld of Al 7075 T-6 can be inside the scope of 425–450 °C on this desired range of temperature voids can be eliminated.

References

1. Shahabuddin, Dwivedi VK, Sharma A (2019) Influence on the tensile properties of AA7075-T6 under different conditions during friction stir welding process. IOP Conf Ser: Mater Sci Eng 691:012001
2. Yoshihiko U, Keiro T, Yasunari T, Hideaki S (2009) Fatigue behaviour of friction stir welded A7075-T6 aluminium alloy in air and 3% NaCl solution. J Japan Weld Soc 27:261–269
3. Varma RR, Ibrahim AB, Mansor MAB (2014) Mechanical properties of the friction stir welded dissimilar aluminium alloy joints. Int J Mech Prod Eng 2:1–5
4. Renjith CR, Raghupathy R, Dhanesh GM (2016) Optimization of process parameters for friction stir lap welding of AA6061-T6 and AA7075-T6 aluminum alloys using Taguchi. Int J Res Technol Stud 3:1–7
5. Ravikumar S, Seshagiri Rao V, Pranesh RV (2015) Effect of process parameters on mechanical properties of friction stir welded dissimilar materials between AA6061-T651andAA7075-T651 Alloys; Prabhuraj P, Rajakumar S, Balasubramanian V, Experimental investigation on mechanical and metallurgical properties of the friction stir welded Aa7075-T651 aluminum alloy joint 9:0974–2115
6. Srinivas Naik L, Hadya B (2020) Evaluation and impacts on mechanical behavior of friction stir welded copper 2200 alloy. In: Narasimham G, Babu A, Reddy S, Dhanasekaran R (eds) Recent trends in mechanical engineering. Lecture notes in mechanical engineering. Springer, Singapore
7. Bayazid SM, Heddad MM, Cayiroglu I (2018) A review on friction stir welding, parameters, microstructure, mechanical properties, post weld heat treatment and defects 2:116–126
8. Muruganandam D, Ravikumar S, Das SL (2010) Mechanical and micro structural behavior of 2024–7075 aluminum alloy plates joined by friction stir welding 9:247–251
9. Shahabuddin, Dwivedi VK, Sharma A (2019) Experimental investigation of the mechanical properties and microstructure of AA 7075-T6 during underwater friction stir welding process. Int J Engi Adv Technol 8:1289–1294
10. Song Y, Yang X, Cui L, Hou X, Shen Z, Yan Xu (2014) Defect features and mechanical properties of friction stir lap welded dissimilar AA2024–AA7075 aluminum alloy sheets. Mater Des 55:9–18
11. Sharma A, Dwivedi VK (2019) A comparative study of micro-structural and mechanical properties of aluminium alloy AA6062 on FSW and TIGW processes. Int J Innovat Technol Explor Eng 8:337–342

Recent Trends in Weldability and Corrosion Behavior of Low Nickel Stainless Steels

Prashant Kumar Pandey, Rajeev Rathi, and Jagesvar Verma

Abstract The current nickel supply deficit and continuous change in its expense brought about getting the consideration of researchers to prompt different options of stainless steel which contains less measure of nickel as its alloying component. Among the different existing stainless steel (SS) grades, austenitic stainless steels (ASS) are used in roughly 60–70% of utilizations, however, ASS requires nickel as its major subsequent. One of the better choices to supplant austenitic stainless steel is ferritic grades, which contain less nickel and have fundamentally the same as execution at a generally lower cost. The current examination is an endeavor to feature the different welding methods (fusion/solid-state) which used most, for joining of these ferritic stainless steel (FSS) grades. The literature affirms that because of the metallurgical issues related to fusion welding processes, solid-state joining processes such as forge welding, pressure welding, ultrasonic welding, and friction stir welding (FSW) are advised to be used for joining of ferritic stainless steels depending their suitability. Moreover, the choice of an appropriate welding process for a particular stainless steel grade is also governed by the corrosion-related aspects, and to reduce the adverse effects of corrosion post welding it is recommended to use pre weld and post-weld heat treatment processes.

Keywords Low Ni SS · FSS · Duplex stainless steel · Welding processes · Corrosion behavior

---

P. K. Pandey (✉) · R. Rathi
School of Mechanical Engineering, Lovely Professional University, Phagwara, Punjab, India
e-mail: prashant.15821@lpu.co.in

J. Verma
Department of Manufacturing Engineering, NIFFT, Ranchi, India

A. K. Dubey et al. (eds.), *Recent Trends in Industrial and Production Engineering*, Lecture Notes in Mechanical Engineering,
https://doi.org/10.1007/978-981-16-3135-1_21

1 Introduction

In the interest of their excellent corrosion resistance, good mechanical properties at elevated temperature, and better weldability, austenitic stainless steel have traditionally been the most widely used grade for different industrial applications [1]. The major constituents of austenitic stainless steels are chromium (Cr) and nickel (Ni). These alloying elements strengthen the resistance to corrosion, the sustainability of steels at higher temperatures, and other properties. Furthermore, Ni-based stainless steels are more scratch-safe, attributable to their natural work solidifying properties. However, the issues associated with nickel availability and its cost variation allows the development of other alternate grades. Figure 1 displays the fluctuation in Nickel's price for the past two years on the London Metal Exchange. It can be easily seen that at the end of 2018 the price was minimum (close to USD 10,400 per tonne) during this period, while it reached a maximum peak to nearly USD 18,500 per tonne in mid of 2019.

Among the other alternatives available for austenitic stainless steel, ferritic stainless steels are well known for their excellent resistance against stress corrosion cracking resistance, oxidation at high temperature, pitting, and crevice corrosion in chloride environments [2]. Besides, less expansion and contraction due to temperature change, exceptional confrontation against oxidation at elevated temperature, and stress corrosion cracking enhance the application of ferritic stainless steels in various industries mainly covering the automobile sector (exhaust pipes, catalytic converters, mufflers, tailpipes, etc.), the petrochemical sector (refineries), food processing,

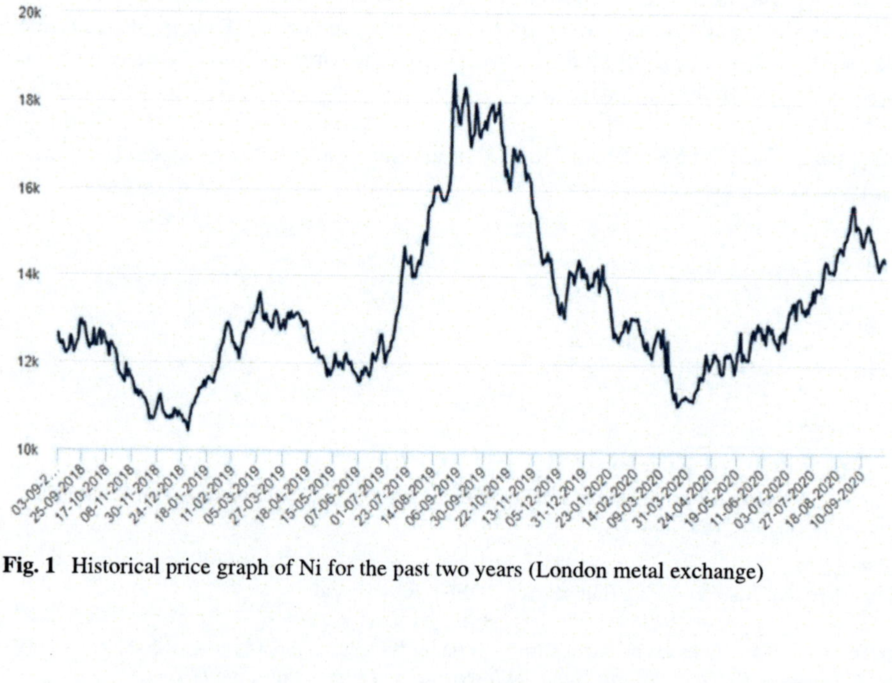

Fig. 1 Historical price graph of Ni for the past two years (London metal exchange)

brewery and wine-making equipment [3]. Reduction in the concentration of interstitial elements such as carbon and nitrogen with the addition of stabilizing elements like molybdenum, titanium, niobium, and others enhances the resistance against corrosion/erosion and it helps ferritic stainless steel grades to be even better than most of the austenitic stainless steel grades [4]. Duplex stainless steel comprised of ferrite and austenite is again one of the alternatives that display its strength and corrosion resistance in numerous aggressive environments. Based on the outstanding performance of these grades of SS, it has been brought to the notice of researchers, manufacturers, and end-users over the past 20 years since development.

Further attention is needed to identify better techniques for joining these materials either similar or dissimilar [5]. These steels can be effortlessly joined by fusion welding processes, while studies show that the heat generated during fusion welding processes leads to coarsening of grains as rapid cooling yields the solidification of molten metal to ferrite phase without any intermediate phase transformation. Thus, the conventional liquid-solid welding processes such as fusion welding decreases the useful mechanical properties such as ductility, impact strength, and corrosion resistance [6].

2 Literature Review

2.1 Literature Search Methodology

This paper involves a unified approach to present innumerable research works done on the joining of similar or dissimilar stainless steel joints with fusion and solid-state welding processes. The literature of the past 10 years is collected (Fig. 2) and an effort

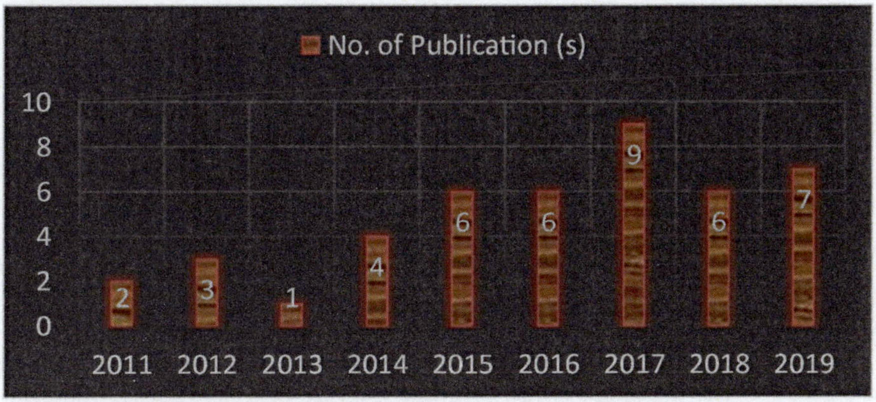

Fig. 2 Number of publication (s) considered for literature (year wise)

has been made to identify the recent advancement in welding of low nickel-based stainless steel.

2.2 *Weldability of Low Ni-based SS*

Among the various solid-state welding processes, friction stir welding proved to be a successful welding process to develop defects free weld of FSS grade 409 utilizing a polycrystalline cubic boron nitride (PCBN) tool. The microstructure analysis displayed a fine-grained disseminated stir zone (SZ). Further, investigation reveals a considerable rise in the existence of the low-angle grain boundary in SZ when compared with a base metal (BM). It has also been detected that increasing the plunging depth decreases the size of the grain and improves hardness [7]. Keeping the friction stir welding tool rotational speed constant and the welding speed as a variable, various tests comprising optical microscopy, electron back-scattered diffraction (EBSD), scanning electron microscopy (SEM), impact test as well as hardness test indicate that the mechanical and microstructure properties of FSS exhibit a substantial change after welding. The major effect is observed on the low-angle grain boundary which increases as a result of fine-grained equiaxed ferrite [8]. Salemi Golezani et al. [9] used a heavy-duty NC machine to perform FSW for joining FSS grade A430 sheets with thickness measures as 2 mm. They studied the impact of traverse and rotating speed of tungsten carbide tools on mechanical and microstructural properties. The use of a brass chamber assisted with water cooling avoids the wear and damage of the tool. Also, protection applied with argon gas provides shielding to counter tool oxidation at elevated temperature. Varying rotational speed and welding speeds were selected. Results disclosed that increasing the welding speed at constant rotational speed decreases ferrite grain size, which verifies active recrystallization existence in the nugget zone. Mehmet et al. [10] analyzed the effects of the various speed of FSW tools including traverse and rotational speed of ferritic stainless steel grade 430. Two samples of grade A430 of the same thickness as 3 mm were butt welded with friction stir welding. It was detected that the greatest values of mechanical properties were attained at a tool traverse speed of 125 mm/min and rotational speed of 1120 /min (Fig. 3). While the tool angle should be kept as 0^0 and a continual pressure force of 3.5 kN needs to be applied.

The rapid cooling done on FSS after its friction stir welding transforms the ferrite grains from coarse to the fine structure since rapid cooling in addition to frictional stirring initiates severe plastic deformation which further develops a high strain. The fine grains allow the joint to exhibit good impact strength/toughness and ductility as well [11, 12].

Laser beam welding (LBW) is also a useful technique to join FSS sheets and to prepare similar/dissimilar joints. With the increase in welding speed, the heat-affected zone (HAZ) becomes narrower and simultaneously the hardness moves up on the higher side. The varying welding speed does not produce any major change in tensile strength, but at a higher welding speed, greater elongation is achieved [13,

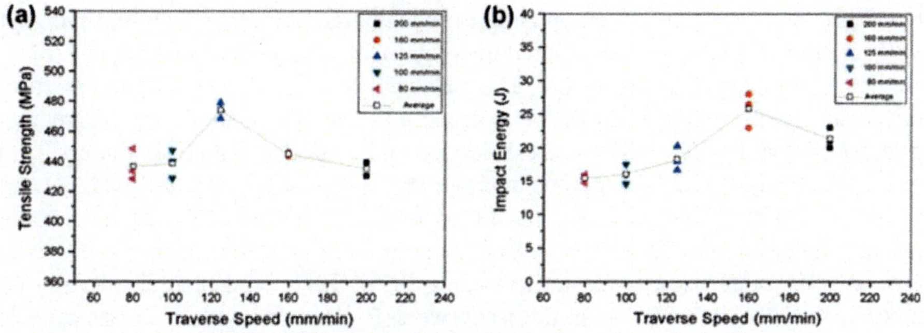

Fig. 3 Effect of tool traverse speed on **a** Tensile strength and, **b** Impact energy [10]

14]. The test results show that the rapid solidification of laser beam welded FSS joints of grade 409 M, reflects excellent bending strength and toughness due to the formation of dendritic grains from the coarse ferrite grains of base metal [15–17].

Once comparing the tensile strength of two different ferritic stainless steel grades, i.e., 444 and 429L, butt welded with the help of gas metal arc welding (GMAW), it has been found that 429L has more tensile strength [18]. Mukherjee et al. [19] experimentally investigated the process response and the microstructure of dissimilar joints made of low Ni ASS grade and 409 M FSS grade. Elemental mapping indicated that the material flow strategy during FSW relies upon the blend of the adopted procedure parameters. Dynamic recrystallization and recuperation are additionally seen in particular dissimilar joints. The FSS displays increasingly serious powerful recrystallization, bringing about an extremely fine microstructure, likely because of the higher stacking energy among the two distinctive grades of steel selected in the present investigation. Gas tungsten arc welding is a versatile process that is frequently used for industrial purposes [20–22]. The FSS grades can also be easily welded by GTAW/TIG and demonstrate better weldability aspects, while the precipitation of carbide and martensite formation is the major issues experienced with this welding technique. These issues should be controlled to hinder unaccepted grain growth in HAZ. It has been found that the intensity of the heat input selected during this welding technique leaves a great effect on grain growth [23– 25]. Narrow HAZ, an increase in microhardness and grain refinement with additional equiaxed crystals can be achieved by adding nitrogen gas in argon-based shielding gas provided with a double layer shielding since nitrogen gets dissolved into the weld pool. Due to microstructure refinement takes the toughness of the welded FSS joint also increases significantly. It has also been observed that the ductile fracture region normally developed very near to the surface of the weld [26, 27]. The pulse mode used for metal transfer enhances the mechanical resistance and metallurgical properties of FSS. This could be attained as this metal transfer mode expressively modifies the composition of the weld metal as compared to spray mode. In spray mode of metal transfer, the stability of the austenite phase gets promoted and at variable heat inputs, the grain structure features considerable enhancement. Pulse mode upgrades hardness as well as impact toughness of weld metal as compared to spray mode at any given heat input [28].

The relationship derived for the resistance spot welding process and microstructure changes on ferritic stainless steel grade 430 show that numerous parameters like the formation of martensite, carbide precipitates at the grain boundary, and grain growth significantly affect the microstructure of heat-affected zone (HAZ) and fusion zone (FZ). It is worth noted that due to the above facts there is an adverse effect on mechanical properties [29, 30]. Bansod et al. [31] investigated the microstructure of FSS joints (similar/dissimilar) using SEM, XRD, and optical microscopy methods and found that in dissimilar joint (when FSS welded with low carbon steel) Widmanstatten ferrite is formed very near to HAZ. With an increase in welding speed, the hardness of the welded zone is improved and at the same time, the HAZ becomes narrower. Varying welding speed does not have any effect on the ultimate strength of the joint but an improved elongation is attained at a higher speed. The test result of corrosion analysis reveals that dissimilar joint possesses better corrosion resistance which is mainly due to the presence of galvanic corrosion. FSS with low Ni percentage displays good strength and better stress corrosion resistance in various corrosive environments [32]. Experimental results show that HAZ of FSS grade 444 welded with electrode E309MoL-16 (AWS) exhibits considerable grain growth as compared to the base metal. Very fine precipitates of needle-shaped were detected in the partly molten zone which can be judged as Laves phase. Also, finely dispersed precipitates were observed in HAZ and this is due to the weld thermal cycle. The X-ray diffraction (XRD) study established the availability of few carbonitrides and chromium nitrides as well as certain secondary phases like chi and sigma [33]. Figure 4 illustrates some of the major constitution diagrams such as Schaeffler, Delong, and WRC-1992 which are useful to predict the microstructure of when dissimilar metal joining is done. However, for stainless steel with higher nitrogen content, the Schaeffler diagram has not been accounted. In this situation, the WRC-92 gives the result that is nearer to reality than those attained using Delong.

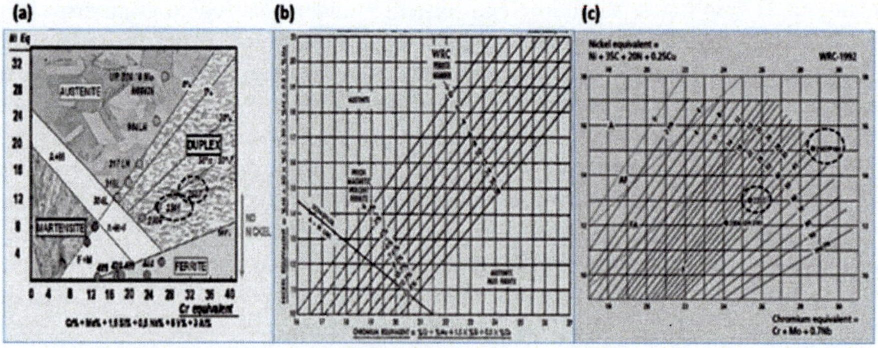

Fig. 4 Constitution diagrams for dissimilar metal joining to predict the weld microstructure **a** Schaeffler, **b** Delong, **c** WRC-1992 [5]

2.3 Corrosion Behavior

Wenyong et al. [34] deliberated the pitting corrosion behavior of FSS grade 444 with the help of electrochemical methods comprising cyclic potentiodynamic polarization, chronoamperometry, and immersion tests besides test to identify corrosion extents. An attempt was made to observe preceding and succeeding corrosion tests, utilizing the scanning electron microscopy (SEM) technique which displays indistinguishable corrosion behavior in all the corroded environments (solutions) subjected to allow the samples for only one minute into the solutions. It is also to be noted that 444 shows improved corrosion confrontation in chloride environments. While in sulphuric acid, it has additional corrosion pits. Khattak et al. [35] analyzed the effect of weld metal interaction on the predisposition of 444 FSS weldment to SCC in hot chloride, by continuous load tests and metallographic investigation. To develop weld joints with unlike chemical configurations in fusion zones, two different filler metals E316L and E309L (both are from austenitic stainless steel domain) were used. The SCC test outcomes revealed the susceptibility of the interface among the FZ and the HAZ. The investigation carried out to find the effect of chloride impurity level and the corrosion mechanism on the susceptibility of FSS and ASS suggests that FSS experiences a substantial mass gain due to the formation of a thick as well as non-protective oxide scale. Furthermore, ASS shows greater resistance to corrosion in high and low chloride salts [36–39].

2.4 Pre Weld Heat Treatment

The analysis was done to examine the effect of annealing on corrosion behavior of FSS and found that annealing may not be able to improve the resistance against crevice corrosion and pitting but it does advance the weldability of the material by hindering carbide precipitations [40–42]. The surface grinding operations are done on FSS to improve the corrosion resistance of the material. These operations hinder the corrosion process at grain boundaries which are located just under the surface layer and also discourage the progression of microcracks when the material is exposed to high loads [43, 44].

3 Discussion of Study

The corrosion-resistant and high-quality materials costs are growing day by day and with this fact, the ferritic steel is a decent option for the utilization in the chloride condition and numerous uranium improvement plants. In these applications welding is the primary procedure for the development of designing structures that debase their properties. In this way, unique consideration is required for the improvement of mechanical properties of FSS during welding which has still a few issues. The use of numerous arc welding processes (like GMAW, SAW, SMAW, etc.) responds in the formation of undesirable precipitates, grain morphologies, and development of residual stresses. While it is shown in earlier works that low-energy welding processes can be preferred but with these, there is an issue of incomplete penetration. In terms of grain refining, intergranular oxidation, and fragility, numerous studies have been conducted on the welding of FSS. Because of these variations, the microstructure and mechanical properties are greatly affected. Among the existing fusion welding processes, GTAW/TIG welding is best suitable to be used for joining of these low Ni stainless steels as this process confirms better weldability as compared to others. Friction stir welding is recommended to be used as a joining technique of various grades of FSS with the thickness of these plates/sheets to be on the lower side. The process parameters should be accurately selected and monitored during the welding. The rapid cooling renovates the ferrite grains from coarse to the fine structure. During the FSW process, argon gas shielding helps to counter tool oxidation at elevated temperatures. Besides, the use of a brass chamber assisted with water cooling avoids the wear and damage of the tool. The literature also illustrates that while performing Laser beam welding (LBW) to join FSS sheets/plates the welding speed should be kept at moderate since it narrows the HAZ is also a useful technique to join FSS sheets and to prepare similar/dissimilar joints. With the increase in welding speed, the heat-affected zone (HAZ) becomes narrower and simultaneously the hardness moves up on the higher side. The research work which is done so far on the corrosion behavior of FSS demonstrates that this steel encompasses better corrosion resistance as compared to most of the current ASS grades. The experimental results for dissimilar weld joints of AISI 304 and AISI 4104 prepared by various welding processes have been summarized in Table 1. It has been observed that the heat affect zone (HAZ) in the GTAW joint is primarily due to the high heat input and low welding speed used during this process. However, the lower heat input and higher welding speed in the case of electron beam welding (EBW) causes the size of the fusion pool to be small and faster cooling rate. As the HAZ is absent in EBW joints that results in better tensile strength. The application of high heat input in GTAW induces the micro-segregation of alloying elements and the formation of chromium depleted zones results in lowering the mechanical properties [45–47].

Table 1 The cumulative mechanical properties of FSS and ASS dissimilar weld joints by FSW, EBW, and TIG [45]

Welding process	Max hardness as weld (HV)	Yield strength (MPa)	%Elongation
Friction welding	305	485.2	18.15
Electron beam welding	400	681	31.97
TIG welding	400	634.5	24.96

4 Conclusion

An attempt has been made to identify welding processes which are suited for joining low Ni SS grades and to study the corrosion behavior of these stainless steel grades in the various corrosive environment. There are various fusion welding processes available that can be utilized for welding of these stainless steel grades, but literature confirms that GTAW/TIG is best suited among those. While using FSW (a solid-state welding process) it is recommended to select appropriate (higher side) plunging depth, as it decreases the grain size and thus improves hardness. Post welding the mechanical and microstructure properties of these grades exhibit an extensive change. Fine-grained equiaxed ferrite contributed to increasing the low-angle grain boundary. The use of an appropriate shielding method (argon gas shielding) helps in getting better weld beads even at higher temperatures.

References

1. Lambert P (2016) Sustainability of metals and alloys in construction. In: Sustainability of construction materials, Woodhead Publishing, pp 105–128
2. Balasubramanian V, Shanmugam K, Lakshminarayanan AK (2009) Effect of autogenous arc welding processes on tensile and impact properties of ferritic stainless steel joints. J Iron Steel Res Int 16:62–68
3. Uematsu Y, Akita M, Nakajima M, Tokaji K (2008) Effect of temperature on high cycle fatigue behavior in 18Cr–2Mo ferritic stainless steel. Int J Fatigue 30:642–648
4. Han J et al (2014) Microstructure and mechanical properties of friction stir welded 18Cr–2Mo ferritic stainless steel thick plate. Mater Des 63:238–246
5. Verma J, VasantraoTaiwade R (2017) Effect of welding processes and conditions on the microstructure, mechanical properties and corrosion resistance of duplex stainless steel weldments—a review. J Manuf Process 25:134–152
6. Cho H-H et al (2011) Microstructural analysis of friction stir welded ferritic stainless steel. Mater Sci Eng 528:2889–2894
7. Alizadeh-Sh M, Marashi SPH, Pouranvari M (2014) Resistance spot welding of AISI 430 ferritic stainless steel: phase transformations and mechanical properties. Mater Des 56:258–263
8. Shunmugasamy VC, Mansoor B, Ayoub G, Hamade R (2018) Friction stir welding of low-carbon AISI 1006 steel: room and high-temperature mechanical properties. J Mater Eng Perform 27(4):1673–1684
9. SalemiGolezania A, Araba SM, Javadi S, Kargar F (2014) The effect of friction stir processing speed ratio on the microstructure and mechanical properties of A 430 ferritic stainless steel. J Adv Mater Process 2:39–48
10. Bilgin MB, Meran C (2012) The effect of tool rotational and traverse speed on friction stir weldability of AISI 430 ferritic stainless steels. Mater Des 33:376–383
11. Lakshminarayanan AK, Balasubramanian V (2010) An assessment of microstructure, hardness, tensile and impact strength of friction stir welded ferritic stainless steel joints. Mater Des 31:4592–4600
12. Kim KH, Bang HS, Kaplan AFH (2017) Joint properties of ultra thin 430M2 ferritic stainless steel sheets by friction stir welding using pinless tool. J Mater Process Technol 243:381–386
13. Fu H, Min J, Zhao H, Yulai Xu, Pengfei Hu, Peng J, Dong H (2019) Improved mechanical properties of aluminum modified ultra-pure 429 ferritic stainless steels after welding. Mater Sci Eng A 749:210–217

14. Mostaan H, Nematzadeh F (2017) Micro laser welding of AISI 430 ferritic stainless steel: mechanical properties, magnetic characterization and texture evolution, pp 1–8
15. Liu LL, Hu SS, Shen JQ, Ma L, Li Y (2019) The effects of average power and pulse duration on weld characteristics, microstructure, mechanical properties and corrosion resistance of a laser welded 21% Cr ferritic stainless steel. Lasers Eng (Old City Publishing) 44
16. Hu S, Pang J, Shen J, Wu W, Liu L (2016) Microstructure, mechanical property and corrosion resistance property of Cr26Mo3. 5 super ferritic stainless joints by P-TIG and laser welding. Transactions of Tianjin University 22(5):451–457
17. Lakshminarayanan AK, Balasubramanian V (2012) Evaluation of microstructure and mechanical properties of laser beam welded AISI 409M grade ferritic stainless steel. J Iron Steel Res Int 19:72–78
18. Verma J, Taiwade RV, Sonkusare R (2017) Effects of austenitic and duplex electrodes on microstructure, mechanical properties, pitting, and galvanic corrosion resistance of ferritic and dual-phase stainless steel dissimilar joints. J Mater Res 32(16):3066–3077
19. Mukherjee M, Saha J, Kanjilal P, Pal TK, Sisodia S (2015) Influence of gas mixtures in GMAW of modified 409M ferritic stainless steel. Weld J 94:101–114
20. Nain SS, Rathi R, Varma B S, Panthangi RK, Kumar A (2020) Machine learning application for pulsating flow through aluminum block. Advances in intelligent manufacturing. Springer, Singapore, pp 189–201
21. Arora H, Bajwa M, Pandey PK, Singh V (2015) Finite element simulation of angular distortion in carbon steel butt welded joint. Int J Appl Eng Res 10(7):17795–17806
22. Kumar R, Bhattacharya A, Bera TK (2015) Mechanical and metallurgical studies in double shielded GMAW of dissimilar stainless steels. Mater Manuf Processes 30(9):1146–1153
23. Khorrami MS, Mostafaei MA, Pouraliakbar H, Kokabi AH (2014) Study on microstructure and mechanical characteristics of low-carbon steel and ferritic stainless steel joints. Mater Sci Eng A 608:35–45
24. Gupta SK, Raja AR, Vashista M, Yusufzai MZK (2018) Effect of heat input on microstructure and mechanical properties in gas metal arc welding of ferritic stainless steel. Mater Res Express 6(3):036516
25. Li H, Xing W, Yu X, Zuo W, Ma L, Dong P, Wang W, Fan G, Lian J, Ding M (2016) Dramatically enhanced impact toughness in welded ultra-ferritic stainless steel by additional nitrogen gas in Ar-based shielding gas. J Mater Res 31(22):3610–3618
26. Keskitalo M, Sundqvist J, Mäntyjärvi K, Powell J, Kaplan AFH (2015) The influence of shielding gas and heat input on the mechanical properties of laser welds in ferritic stainless steel. Phys Procedia 78:222–229
27. Mukherjee M, Dutta A, Kanjilal P, Pal TK, Sisodia S (2015) Enhancement of microstructural and mechanical properties by pulse mode of metal transfer in welded modified ferritic stainless steel. ISIJ Int 55(7):1439–1447
28. Sharma A, Pandey P, Singh H, Arora H (2017) Friction stir welding of aluminium alloy with a distinct material sandwiched
29. Pouranvari M, Marashi SPH, Alizadeh-Sh M (2015) Welding metallurgy of dissimilar AISI 430/DQSK steels resistance spot welds. Weld J 94(6)
30. Zhang Y, Guo J, Li Y, Luo Z, Zhang X (2019) A comparative study between the mechanical and microstructural properties of resistance spot welding joints among ferritic AISI 430 and austenitic AISI 304 stainless steel. J Mater Res Technol
31. Bansod AV, Patil AP, Verma J, Shukla (2019) Microstructure, mechanical and electrochemical evaluation of dissimilar low Ni SS and 304 SS using different filler materials. Mater Res 22(1)
32. Onsekiz M, Altunpak Y (2017) Effect of electrode materials type on resistance spot welding of AISI 430 ferritic stainless steel. Int J Eng Res Africa Trans Tech Pub 31:53–58
33. Bensaid N, Hadji M, Badji R, Benlamnouar MF, Saadi T, Senouci S (2019) Microstructure and mechanical behavior of AISI 430 FSS welds produced with different elemental metal powder addition. In Solid State Phenomena Trans Tech Publications 297:195–203
34. Wu W, Shengsun Hu, Shen J (2015) Microstructure, mechanical properties and corrosion behavior of laser welded dissimilar joints between ferritic stainless steel and carbon steel. Mater Des 65:855–861

35. Khattak MA, Zaman S, Tamin MN, Badshah S, Mushtaq S, Omran AAB (2017) Effect of welding phenomenon on the microstructure and mechanical properties of ferritic stainless steel–a review. J Adv Res Mater Sci 32:13–31
36. Tripathi AM, Sharma A, Patra BB, Pandey PK, Chand R, Rana G (2017) A review on friction stir welding of aluminium alloys: mechanical properties and metallurgical observations. Int J Mech Eng Technol 8(7)
37. Potgieter JH, Adams FV, Maledi N, Van Der Merwe J, Olubambi PA (2012) Corrosion resistance of type 444 ferritic stainless steel in acidic chloride media. J Chem Mater Sci 2:37–48
38. Antunes PD, Correa EO, Barbosa RP, Silva EM, Padilha AF, Guimaraes PM (2013) Effect of weld metal chemistry on stress corrosion cracking behavior of AISI 444 ferritic stainless steel weldments in boiling chloride solution. Mater Corros 64(5):415–421
39. Kumar R, Rathi R, Sharma S (2020) Recent innovation on synthesis methods of graphene-based composites. Adv Mater Sci Engi 11
40. Dorcheh AS, Galetz MC, Durham RN (2016) Effect of chloride contents on the corrosion behavior of ferritic and austenitic steels in molten solar salts. In: International corrosion conference proceedings, NACE International, p 1
41. Zhou N, Pettersson R, Schönning M, Peng RL (2018) Influence of surface grinding on corrosion behavior of ferritic stainless steels in boiling magnesium chloride solution. Mater Corros 69(11):1560–1571
42. Arivazhagan N, Singh S, Prakash S, Reddy GM (2011) Investigation on AISI 304 austenitic stainless steel to AISI 4140 low alloy steel dissimilar joints by gas tungsten arc, electron beam and friction welding. Mater Des 32(5):3036–3050
43. Prithiraj A, Otunniyi IO, Osifo P, van Der Merwe J (2019) Corrosion behaviour of stainless and carbon steels exposed to sulphate–reducing bacteria from industrial heat exchangers. Engi Fail Anal 104:977–986
44. Borgioli F, Galvanetto E, Bacci T (2018) Corrosion behaviour of low temperature nitrided nickel-free, AISI 200 and AISI 300 series austenitic stainless steels in NaCl solution. Corros Sci 136:352–365
45. Xie Yi, Guo S, Leong A, Zhang J, Zhu Y (2017) Corrosion behaviour of stainless steel exposed to highly concentrated chloride solutions. Corros Eng Sci Technol 52(4):283–293
46. Bai G, Lu S, Li D, Li Y (2016) Influences of niobium and solution treatment temperature on pitting corrosion behaviour of stabilised austenitic stainless steels. Corros Sci 108:111–124
47. Bitondo C, Bossio A, Monetta T, Curioni M, Bellucci F (2014) The effect of annealing on the corrosion behaviour of 444 stainless steel for drinking water applications. Corros Sci 87:6–10

Reliability Analysis—A Critical Review

Janender Kumar, Suneev Anil Bansal, and Munish Mehta

Abstract Present article discusses the problem of lesser availability of the operating systems in process industries. Complex systems, continuous breakdowns and unplanned maintenance strategies are the major reasons for poor performance of the plants. The concept of reliability plays a significant role in improving the system availability. Uninterrupted service and long-run availability are the basic needs of complex systems as in the sugar industry, thermal power plants, milk industry, mining, petroleum industries, etc. Different approaches are utilized by researchers in various fields to check the performance of the operating equipment. These approaches are genetic algorithm (GA), Petri nets (PN), fault tree analysis (FTA), supplementary variable technique (SVT), fuzzy Lambda-Tau technique, reliability, availability and maintainability (RAM) analysis, reliability, availability, maintainability, dependability (RAMD) analysis, failure mode and effect analysis (FMEA) and degradation modelling techniques, etc. The improvement has been seen in the performance of the systems based on mathematical data using the above techniques. Article also proposes new areas where research can be carried out.

Keywords Reliability · Availability · Maintainability · Operating system · Failure · Mathematical models · Process industry

1 Introduction

In manufacturing sectors due to high production targets and continuously increased demand, the industries are growing rapidly and simultaneously, the operating equipment in these industries is at high risk. The probability of failure is more. Asset reliability is a very significant feature of quality and reliable performance across its

J. Kumar (✉) · S. A. Bansal
Department of Mechanical Engineering, MAIT, Maharaja Agrasen University, Solan, Himachal Pradesh 174103, India

M. Mehta
School of Mechanical Engineering, Lovely Professional University, Phagwara, Punjab, India

A. K. Dubey et al. (eds.), *Recent Trends in Industrial and Production Engineering*, Lecture Notes in Mechanical Engineering,
https://doi.org/10.1007/978-981-16-3135-1_22

anticipated life cycle. In the process industry, sudden or unexpected failure of any system or component may result in severe interruption or accident in the manufacturing unit. It leads to a direct loss to the concerned industry. Reliability is one of the tools that is used to check the consistency of any operational system. However, availability is the probability that a particular component/system when used under given conditions will perform satisfactorily when required. Maintainability is the likelihood that during the specified time when maintenance action is carried out following the prescribed protocol, an object can be restored to its operating performance. Then, it becomes very important to monitor maintenance strategies. The performance of plant equipment can be enhanced with reliability, availability and maintainability analysis [6]. Unreliability and unavailability of the operating system are dangerous for plant efficiency. Research on portent and asset life prediction in the area of engineering asset management has increased in recent years. The assets degrade over time or deteriorate due to some factors. Degradation means a reduction in lifespan, reliability and performance of assets [9]. Degradation models play a crucial role in reliability assessment. The reliability of the system can be evaluated with reliability prediction which is based on degradation modelling. In reliability analysis, degradation models can be classified as shown in Fig. 1.

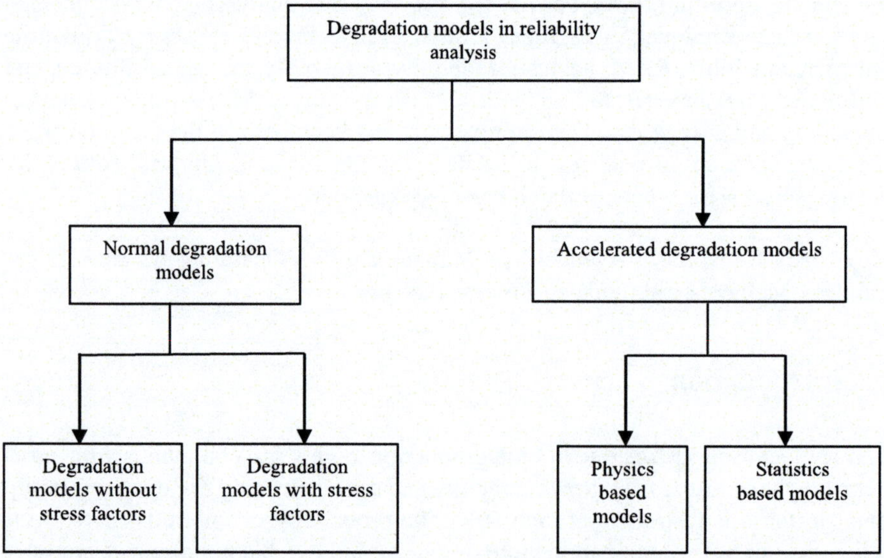

Fig. 1 Classification of degradation models in reliability analysis

2 Literature Review

Several studies have been performed to access the importance and existence of the reliability concept in industries. Results from these studies have improved output for the process industries. If the system arrangement is in series, the failure of one unit stopped the whole operating system, and in parallel, if comes in failure state, it reduces the capacity of the plant. So, for the highest growth and survival, it becomes important that the operating system should run without an obstacle. The various research papers are critically reviewed and compiled to give an overview of reliability analysis.

Gupta et al. (2008) evaluated the feed water unit of a thermal power plant. Mathematical model was developed with a probabilistic approach. Differential equations were solved with steady-state analysis's principles. Finally, the availability of the system equation was developed. Based on available data, maintenance priorities were decided for the feed water pump, condenser and boiler of a feed water unit system. The priority was given to feed water system based on its failure and repair rate as it was higher than other systems [1].

Mehta et al. (2012) performed a study in the milk industry. The performance of the butter oil production system is analysed by Markov birth and death process. The system was categorized into three states, i.e. full working capacity, reduced capacity (parallel system) and failed state. The comparative results between three subsystems showed that the heater subsystem contributes more in decreasing availability as failure rate increased, i.e. 14.42%, and if the repair rate increased, then availability increased 13.36%. Then priority was given to the heater subsystem from the maintenance point of view [2].

Garg et al. (2010) performed the study in the screw plant industry for reliability and availability. After calculating the availability equations of the system, the decision matrix was made with the help of repair and failure rate data obtained from the industry. Then availability w.r.t. repair and failure rate was analysed from the decision matrix table of data. General algorithm was utilized to check the optimal value of availability [3].

Sharma and Vishwakarma (2014) discussed the complex system of the process industry. As for the better performance of the unit, all the operating equipment must work consistently and for a long time. So, the reliability, availability and maintainability of this equipment have very much importance. The reliability of the feeding system of the unit consisting of cutting, crushing, bagasse-carrying and heat-generating subsystems was considered for analysis. Markov birth and death process was used for analysis. With the known values of failure rate and repair rate of all subsystems, availability was calculated, and based on those maintenance priorities, preferences were assigned. Further, an optimality test was performed with a genetic algorithm technique for evaluating system performance [4].

Mehta et al. (2018) investigated the reliability analysis of an industrial unit. It was stated that breakdowns cannot be eliminated but can controlled up to some extent. This can be only done with a better maintenance policy/strategy. In this research paper, a mathematical model of a butter oil manufacturing unit was developed. From the state transition diagram of the system, differential equations were developed by applying supplementary variable technique (SVT) technique. These equations were then solved by Lagrange's method, and availability was calculated using Runge–Kutta fourth-order method. Finally, based on the failure rate, the priority was given to the clarifier subsystem, whereas on behalf of the repair rate, it was given to the heater subsystem [5].

Sharma and Kumar (2008) stated that the performance of equipment can be maximized by reliability, availability and maintainability (RAM) analysis. To optimize the reliability and maximize the equipment availability, it was important to achieve minimum failure or increase the mean time between failures (MTBF). The Markovian approach has been utilized to know the behaviour of the system. At the end of the study, the result proved helpful in maintaining the strategies of maintenance programs [6].

Aggarwal et al. (2016) carried the reliability and availability analysis of a skim milk power industry. The six subsystems like cream separator, chiller, packaging system, drying chamber, evaporator and pasteurizer were analysed. These subsystems are arranged in series and parallel way of configuration. Markov birth and death process was implemented. Chapman–Kolmogorov differential equations were derived. These equations were then solved using Runge–Kutta fourth-order method. The result shows that cream separator puts more impact on availability as its failure rate is large and then even MTBF. So, proper attention should be given to this subsystem to enhance the performance of the system [7].

Aggarwal et al. (2015) proposed the research of a skim milk powder system of a dairy plant. It had six subsystems out of which cream separator and chiller subsystems were critical from the viewpoint of maintenance. Reliability, availability, maintainability and dependability (RAMD) analysis was performed which set the priorities/policies of maintenance. The results that obtained from RAMD analysis helped the management in deciding maintenance priorities of the plant [8].

Gorjan et al. (2009) included the health and life prediction of assets for reliability analysis in the industry. It was stated that in recent years, research on assets health and lifespan prediction has increased in the field of engineering asset management (EAT). It has introduced a new term of degradation. Basically, degradation means a reduction in reliability, performance, health and lifespan of assets. Degradation data play an important role as it provides significant information than failure time data for analysing the reliability and in predicting the lifespan of the assets. Degradation modelling techniques play an important role in the field of reliability analysis. These models have sufficient potential in the prediction of the asset's health and reliability [9].

Yadu et al. (2016) considered the automated band saw cutting machine for analysing its reliability and life cycle cost. With an improved design, it was observed that system reliability increased by 15.85% and life cycle cost reduced by 22.09%. MTBF increased from 15,000 h to 20,000 h [10].

Yang et al. (1998) compared two methods for assessment of reliability, i.e. degradation modelling path and traditional life test. Degradation means a reduction in reliability, performance, health and lifespan of assets. The product may degrade or deteriorate over time. Hence, reliability declines. In this paper, the severe critical values technique was used for degradation. Higher the rank of severe critical value, shorter will be the lifespan of the product, whereas traditionally life tests only tell the time to failure for reliability analysis. It was observed that the new method was much cost-effective and accurate [11].

Gerbecs (2010) performed the study for validation of existing maintenance policies, reliability, availability and operability of pressure regulating installation. The high-pressure natural gas has transmission through a pipeline network to the consumer. In this paper, a case study was done for this crucial pressure regulating installation system. The important things learnt through a case study were as follows: Due to the reverse flow of natural gas at the consumer end, it brings external transient pressure disturbance in the pipeline. It reduced the overall reliability of pressure regulating installation. The problem at the consumer end should be removed in this case. Secondly, the research work declared that existing countermeasures against back pressure and over pressure are not sufficient for handling the transient pressure disturbance. Again, it left some work for further research that what should be the countermeasure to solve it [12].

Ding et al. (2011) conducted the study for cutting tool wear degradation states. The study was conducted on a computer numerical control (CNC) turning centre. It was stated that tool wear or blunt tool not only affects the dimensional accuracy but also degrades the quality of surface finish. In the current paper, the vibration signals of the tool holder were observed to see the degradation of tool wear. The proportional hazards model was used for condition monitoring and also used in tool wear monitoring on CNC lathe [13].

Furuly et al. (2013) investigated the mining industry in Norway. Due to its complex nature, its reliability and availability of the system need more attention. In the present study, the risk rate of stacker belt which was a transportation system and exposed to the environment was analysed. It had to work in different climate conditions. That is why the need arises to consider this system on a priority basis in the maintenance plan. It was concluded that the stacker belt gets more affected in the winter season in terms of hazard rate (four times) than the rest of the year. So, maintenance policy needs to be changed in the winter season which may increase the MTBF [14].

Barends et al. (2012) showed the influence of failure mode and effect analysis approach in the pharmaceutical industry. The consistency of the results obtained with traditional FMEA has been re-evaluated with the probabilistic modification of FMEA. Studies show that the result based on RPN values alone can be misleading. So, scores used to calculate RPN should be considered separately for analysis. In this paper, a new approach which was the modified version of the older FMEA approach has been used. The new approach does not take more time than older FMEA. The major disadvantage of this method was the precision of estimated frequencies of undetected failure modes values which were very large, i.e. in digits which gives the feeling of inaccuracy. So, it represents that a new approach requires additional expertise and training among the FMEA team [15].

Rahbi et al. (2017) analysed the system behaviour in aluminium industry plant with semi-Markov process and regenerative point techniques. The paper deals with the subsystem named butt shot blast station whose performance affects the functioning of the plant. Six years of maintenance data were collected for research purposes. The following measures have been calculated as follows: MTBF-58 h, availability—0.8467, busy period of repairman—0.15 and expected number of visits by repairman for repair—0.16. [16].

Yuan et al. (2011) illustrated the standby redundancy technique which was of two types: cold and warm standby. It has been utilized to enhance system reliability and availability. Cold standby means that redundant units cannot fail while they remain idle when compared to warm standby. In this paper, the warm standby system was analysed which has two dissimilar units and one repairman. Working and repair time distribution of both units were exponential. Unit one was mostly used. A transfer switch was unreliable in the system, and its failure means the complete system fails immediately [17].

Yunusa et al. (2017) investigated the unit of a cement manufacturing industry. Rotary kiln, a very crucial asset of a fully automatic-type cement process plant, has been considered for analysis. The downtime associated with that asset was the huge loss to the industry. The chronic rotary kiln refractory brick failure was analysed by two famous risk analysis techniques. These are fault tree analysis and reliability block diagram. It was concluded that the occurrence of failure of rotary kiln refractory brick can be eliminated or reduced by following the recommendations of the investigation team [18].

Piadeh et al. (2018) researched hybrid systems of advanced treatment units (ATUs) in the parks of Iran. These units used to treat wastewater and made it usable. The new framework for reliability assessment has been proposed. The two techniques fault tree analysis and combined event tree analysis were utilized. The paper represents that the maximum probability of failure was due to osmosis unit with 30%, and the most reliable hybrid system has 74.82% reliability. The proper data and experience help in reducing the risk of capital investment and so on [19].

Badida et al. (2019) focused the research in oil and gas sectors within the country which plays a crucial role in the economic development, as the oil or gas was transmitted within the country and outside the country through pipelines. The risk factor was very high because of natural disaster, and any reason for the failure of these pipes can put a harmful impact on the environment and their survivors. Due to the absence of historical data of these pipelines, the study has been conducted to find out the probabilities of failure. So that risk could be avoided. Fuzzy fault tree analysis technique is implemented for analysing the probability of these severe failures of this sector which affected by the combination of multiple natural hazards. The result obtained helped in making the decision related to risk management of oil and gas pipelines caused by single as well as multiple natural hazards [20].

Manoharan et al. (2019) investigated the study for failure data analysis in a steel rolling mill. The product of which was thermo-mechanically treated (TMT) rod. A framework has been designed for evaluation of failure mode and reliability improvement by design modification to reduce the idle time which exists due to breakdowns and also cost associated with this. FTA approach is utilized for risk assessment. Then according to study outcomes, the rotary shearing machine is introduced to increase performance. The reliability of the furnace section increased by 17% and continuous mill by 15% in TMT rod manufacturing steel rolling mill [21].

Uniyal et al. (2020) presented the article on optimization techniques. Various researchers have utilized many techniques to achieve optimality in optimization problems. In this paper, an overview of nature-inspired meta-heuristics techniques has been discussed which was gaining popularity these days. These were ant colony optimization, particle swarm optimization and grey wolf optimization [22].

Gowid et al. (2014) proposed a standby unit for a propane pre-cooled mixed refrigerant liquefaction plant to increase reliability and minimize the maintenance cost. Time-dependent Markov approach was used to analyse the failure rate of the liquefaction plant. Isograph software and MATLAB were also used for optimizing the calculations. The result showed that unavailability reduced from 14.75 to 2.9% of total operating hours and profit increased by 180 million USD [23].

Du et al. (2017) developed a new model named Markov repairable system with history-dependent up and down states including neglected failures. The failure which has been recovered in a very short period and does not affect the performance of the system was treated as a neglected failure. In the study, the theory of aggregated stochastic processes was used to develop this new model [24].

Zhang et al. (2017) proposed a four-step framework for reliability assessment of a system which was in service. In this type of system and its components, information was available in inspection reports. An evidence theory-based framework has been utilized for assessing the system reliability whose data were available in text form. Naive Bayes classification method was used to treat the text data. It counts the frequency of positive and negative information for the component condition. Therefore, multiple data have been collected which describes the condition of the component. Then interval basic probability assignments with evidence network have been utilized to obtain overall system reliability [25].

Zhang et al. (2013) analysed the reliability of a modular converter invented by Hjort. It is an electronics device that converts electrical signals from one level to another and plays a crucial role in the wind turbine. It has the highest failure rate in wind turbines. This type of architecture with six feasible converter system models has been studied. Markov modelling approach was employed for assessing the reliability of these models. Finally, after comparison and perception, guidance is provided to convertor designers to reduce the risk associated with its design [26].

Dharmaraja et al. (2016) studied the reliability and survivability of the vehicle ad hoc network. This technology provides communication between moving vehicles through mobile Internet. The performance of this network depends upon the hardware and channel availability for the vehicle-to-vehicle and vehicle-to-equipment communication. The connectivity and message lost due to unreliable hardware and channel availability were the major areas of concern during analysing reliability and survivability. Reliability block diagram and Markov chains techniques were used for both types of communications for analysis. The study concluded with remarks that loss of communication was due to hardware components failure [27].

Rashid et al. (2015) highlighted the lubrication system of a helicopter. Main gearbox unit was considered because the loss of oil was increasing the friction between its components. This leads to failure of gearbox and helicopter itself. The reliability analysis of the main gearbox has been done to find out the reason for loss. The influence diagram approach was used to find out the possible factors indicated that lubricant viscosity, design features and maintenance frequency could be the reasons for this failure. Based on rank, it was identified that design was not up to the mark [28].

Kim et al. (2017) proposed an advanced reliability redundancy allocation problem which helps in allocating the optimal redundancy for cold standby system. Study revealed that the cold standby redundancy system was more useful to improve the reliability of the system than hot, warm standby systems. A new reliability model, parallel genetic algorithm, was suggested for solving reliability redundancy allocation problem [29].

Kong et al. (2015) focused the research on the redundancy allocation problem. In general, redundancy allocation for each subsystem is predefined and fixed. In the current study, the focus was on redundancy allocation problem with multiple strategy choices in which additional decision variable for redundancy has been considered. The purpose was to maximize system reliability. Mathematical analysis has been done for series–parallel system. The outcomes proved that a simplified version of particle swarm optimization was a good alternative for solving the redundancy allocation problem with multiple strategy choices [30].

Byun et al. (2017) analysed the reliability growth of a complex system which consists of the number of components and subsystems. In this paper, a complex system K-out of-N: G was considered for providing a perfect level of redundancy strategy in the system. It means K-components for survival are required from the entire system containing N-components. A new approach was proposed to test the growth in reliability for the K-out of-N system [31].

Endrenyi et al. (2006) discussed ageing, maintenance programs and reliability of the devices/equipment. Ageing means each equipment has a certain life; as time passes, it brings the device near to failure. As the deterioration increases, the asset's condition also deteriorates. The life of the equipment can be prolonged by doing timely maintenance. Analysis shows that maintenance extends the life of assets. But different maintenance policies have separate results. As it also increases the cost, therefore, this should be balanced by the gain of reliability of the system. During reliability assessment, degradation and cost analysis should be properly analysed [32].

Liu et al. (2011) suggested that the reliability engineering concept should be considered or embedded during the product development stage. Due to competition in the global market, it becomes necessary to launch new products frequently. So, the research and development department of any industry might focus on the reliability of components from design to the production stage. It may be very risky for low-level industries if project collapses at final stage. FMEA approach was implemented, and drastic results were obtained. New product defects decreased by 10% and satisfaction level of the consumer rose from 85 to 90% [33].

Schoenig (2006) performed the reliability analysis in the car industry where the hybrid system was developed for safety features. Generally, the design process was improved at an early stage to reduce any fault. A quantitative analysis method was proposed to estimate the occurrence of rare events which were identified beforehand. Markov approach was used to identify the results [34].

Kim et al. (2006) analysed the situation assessment of nuclear power plant's operators which means during the abnormal condition in the plant, operator tries to recognize what is happening in nuclear power plant. Operator might understand the current situation according to his mental level. A new analytical model for a better understanding of NPP operator has been developed which is based on Bayesian interface. This model has been compared with the existing model. The proposed model has been observed more efficient and reasonable as it provides a quantitative description of human operators in respect of situation assessment. Still proposed model has certain limitations which needs certain modifications [35].

Rajpal et al. (2006) proposed an artificial neural network for optimizing the complex repairable system. The model was applied to the helicopter transportation facility. It was an effective and reliable algorithm. Developed ANN model could assess the system performance based on past data. The developed network was found very valuable for a complex repairable system [36].

Davis et al. (2007) analysed the failure rate in buried polyvinyl chloride pipelines used for water transmission. The study shows that the failure rate in PVC pipelines was lesser as compared to traditional pipeline materials like cast iron and asbestos cement. The two-parameter Weibull distribution model was applied, and validation of the model was done by comparing the expected failure rates with aggregated failure data obtained from the UK water industry research database [37].

Zhang (2008) studied the simple repairable system with a delayed repair. In earlier studies, an assumption has been taken that after repairing, the system was as good as new. But here it is assumed that the system was not as good as new after repair. That is why the model appears realistic. With this assumption, the important reliability indices like system availability, mean time to first failure and rate of occurrence of failure are derived through supplementary variable technique and geometrical process. Optimal replacement policy for minimizing the cost per unit time was developed, and uniqueness of this policy has been proved numerically [38].

Haarla et al. (2008) evaluated the reliability of the power transmission system. A reliability model was proposed for power system precisely which enables to do complete estimation for system breakdowns. The developed method applies to transmission grids in reality [39].

Leicester et al. (2008) developed a reliability model to predict the risk of a termite attack on a house. The information was collected from the national database and on a survey of expert opinion for developing a model. The combined information provides a method which predicts the attacking of termite on house [40].

Galante et al. (2009) worked on improving the maintenance policy. The study proposed a preventive maintenance policy for the stationary model only. An exact algorithm was proposed to find out the set of components that could be maintained jointly rather separately to minimize the maintenance cost. The case was studied for ship maintenance [41].

Garg et al. (2010) analysed the reliability analysis of combed sliver production system of a yarn production unit. A mathematical model was formulated using a supplementary variable technique to increase production rate. Finally, it was concluded that unilap and carding machines have much impact on system availability. So, the repair rate of these must be improved to get maximum availability [42].

Kumar et al. (2020) investigated the performance behaviour with respect to availability of a plywood manufacturing plant with a licensed Petri nets software package named GRIF—predicates Petri module. The Veneer layup system was selected for study. It has three subsystems—glue spreader, heat press and trimming saw. It was concluded that a decrease in mean time between failure and reduction in repair rate of heat press subsystem reduced the system availability by 37.42% which was the highest from all subsystems. The outcome indicates that heat press subsystem must be on the top-most priority during maintenance [43].

3 Discussion and Prospective

The growth of any process industry depends upon the availability of its assets and maintenance strategy. Hence, to get maximum output from operating systems, it is very much important that these must be looked after minutely so that the level of failure and repair rate could be minimized. Reliability analysis is a vital part of any industry. It is performed in many process industries to improve performance of the system. A number of different techniques have been utilized to enhance the performance of the system. The availability of operating systems increased, and tangibly, the efficiency of the overall plant gets increased too [10, 33]. From the literature review, it has been observed that still a lot of work is required in certain industries. So future research work can be carried out in the cement industry, textile industry, fertilizer plant and oil refineries, etc., where the behaviour of the system can be analysed. The ant colony optimization, particle swarm optimization (nature-inspired techniques) [22] and Petri nets approach can be utilized for reliability optimization of complex systems in these plants.

4 Conclusions

Present report critically analyses various studies on reliability analysis, particularly in process industry. Following conclusions can be drawn from the present study:

1. Reliability analysis concept helps to compute the system availability and also increasing the mean time between failures, as different research papers are the evidence of its success.

2. The literature revealed that the number of reliability techniques has been utilized in different industries, and positive results are observed in terms of system availability.
3. It provides the platform for the treatment of operating systems through maintenance. Different policies or strategies can be planned to increase the availability of critical components.
4. Reliability analysis helps in identifying the availability of different subsystems. Based upon that availability data, priorities of the subsystem's equipment can be fixed for maintenance. The purpose is to escalate the life of assets.
5. Reliability analysis helps in reducing the cost of replacement of components because it is not possible every time to replace them.
6. Degradation of equipment reduces the reliability of the system. Reliability analysis is done based on failure and repair data, whereas degradation data help us in predicting the lifespan and health of equipment.
7. In recent years, the degradation concept has been rapidly spread in process industries as it can increase the reliability of the plant.
8. Nature-inspired techniques are gaining popularity these days like ant colony optimization, particle swarm optimization and grey wolf optimization.

References

1. Gupta P, Tiwari PC, Sharma AK (2008) Performance modeling and decision support system of feed water unit. South African J Industr Eng 19(2):125–134. ISSN: 2224-7890
2. Mehta M, Kumar B, Kumar S (2012) Performance evaluation of butteroil production system of a milk plant. In: International conference on advancement and futuristic trends in mechanical engineering. PTU, Jalandhar, pp 230–235
3. Garg D, Kumar K (2010) Availability optimization for screw plant based on Genetic Algorithm. Int J Eng Sci Technol 2(4):658–668
4. Sharma SP, Vishwakarma Y (2014) Application of Markov process in performance analysis of feeding system of sugar industry. J Industr Mathe 2014:1–9. https://doi.org/10.1155/2014/593176
5. Mehta M, Singh J, Sharma S (2018) Availability analysis of an industrial system using supplementary variable technique. Jordan J Mech Industr Eng 12(4):245–251. ISSN: 1995-6665
6. Sharma RK, Kumar S (2008) Performance modeling in critical engineering systems using RAM analysis. Reliab Eng Syst Saf 93(6):913–919. https://doi.org/10.1016/j.ress.2007.03.039
7. Aggarwal AK, Kumar S, Singh V (2016) Reliability and availability analysis of the serial processes in skim milk powder system of a dairy plant: a case study. Int J Industr Syst 22(1):36–62
8. Aggarwal AK, Kumar S, Singh V (2015) Performance modeling of the skim milk powder production system of a dairy plant using RAMD analysis. Int J Qual Reliab Manage 32(2):167–181. https://doi.org/10.1108/IJQRM-01-2014-0007
9. Gorjian N, Ma L, Mittinty M, Yarlagadda P, Sun Y (2009) A review on degradation models in reliability analysis. In: Proceedings of the 4th World Congress on engineering asset management, pp 369–384, Greece
10. Waghmode LY, Patil RB (2016) Reliability analysis and life cycle cost optimization: a case study from Indian industry. Int J Qual Reliab Manage 33(3):414–429. https://doi.org/10.1108/IJQRM-11-2014-0184.

11. Yang K, Yang G (1998) Degradation reliability assessment using severe critical values. Int J Reliab Qual Safety Eng 5(1):85–95
12. Gerbec M (2010) A reliability analysis of a natural-gas pressure-regulating installation. Reliab Eng Syst Saf 95(11):1154–1163. https://doi.org/10.1016/j.ress.2010.06.022
13. Ding F, He Z (2011) Cutting tool wear monitoring for reliability analysis using proportional hazards model. Int J Adv Manuf Technol 57(5–8):565–574. https://doi.org/10.1007/s00170-011-3316-4.
14. Furuly S, Barabadi A, Barabady J (2013) Reliability analysis of mining equipment considering operational environments-a case study. Int J Performab Eng 9:287–294
15. Barends DM, Oldenhof MT, Vredenbregt MJ, Nauta MJ (2012) Risk analysis of analytical validations by probabilistic modification of FMEA. J Pharm Biomed Anal 64–65, 82–86. https://doi.org/10.1016/j.jpba.2012.02.009
16. Rahbi YAL, Rizwan SM, Alkali BM, Cowel A, Taneja G (2017) Reliability analysis of a subsystem in aluminum industry plant. In: 6th international conference on reliability, infocom technologies and optimization (ICRITO), pp 199–203. AU, Noida
17. Yuan L, Meng XY (2011) Reliability analysis of a warm standby repairable system with priority in use. Appl Math Modell 35(9):4295–4303. https://doi.org/10.1016/j.apm.2011.03.002
18. Yunusa-Kaltungo A, Kermani MM, Labib A (2017) Investigation of critical failures using root cause analysis methods: case study of ASH cement PLC engineering. Fail Anal 73:25–45. https://doi.org/10.1016/j.engfailanal.2016.11.016
19. Piadeh F, Ahmadi M, Behzadian K (2018) Reliability assessment for hybrid systems of advanced treatment units of industrial wastewater reuse using combined event tree and fuzzy fault tree analyses. J Clean Prod 201:958–973. https://doi.org/10.1016/j.jclepro.2018.08.052
20. Badida P, Balasubramaniam Y, Jayaprakash J (2019) Risk evaluation of oil and natural gas pipelines due to natural hazards using fuzzy fault tree analysis. J Nat Gas Sci Eng 66:284–292. https://doi.org/10.1016/j.jngse.2019.04.010
21. Manoharan P, Selvam Dennsion M, Ganesan V, Palanisamy S (2019) Reliability enhancement of steel rolling mill using fault tree analysis. U.P.B. Sci Bull Series D 81(1):165–178
22. Uniyal N, Pant S, Kumar A (2020) An overview of few nature inspired optimization techniques and its reliability applications. Int J Math Eng Manage Sci 5(4):732–743. https://doi.org/10.33889/ijmems.2020.5.4.058
23. Gowid S, Dixon R, Ghani S (2014) Optimization of reliability and maintenance of liquefaction system on FLNG terminals using Markov modelling. Int J Qual Reliab Manage 31(3):293–310. https://doi.org/10.1108/IJQRM-12-2012-0156
24. DuS ZengZ, CuiL KR (2017) Reliability analysis of Markov history-dependent repairable systems with neglected failures. Reliab Eng Syst Saf 159:134–142. https://doi.org/10.1016/j.ress.2016.10.030
25. Zhang X, Mahadevan S, Deng X (2017) Reliability analysis with linguistic data: an evidential network approach. Reliab Eng Syst Saf 162:111–121. https://doi.org/10.1016/j.ress.2017.01.009
26. Zhang CW, Zhang T, Chen N, Jin T (2013) Reliability modeling and analysis for a novel design of modular converter system of wind turbines. Reliab Eng Syst Safety 111:86–94. https://doi.org/10.1016/j.ress.2012.10.005
27. Dharmaraja S, Vinayak R, Trivedi KS (2016) Reliability and survivability of vehicular ad hoc networks: an analytical approach. Reliab Eng Syst Saf 153:28–38. https://doi.org/10.1016/j.ress.2016.04.004
28. Rashid HSJ (2015) Reliability model for helicopter main gearbox lubrication system using influence diagrams. Reliab Eng Syst Safety 139:50–57. https://doi.org/10.1016/j.ress.2015.01.021
29. Kim H, Kim P (2017) Reliability–redundancy allocation problem considering optimal redundancy strategy using parallel genetic algorithm. Reliab Eng Syst Saf 159:153–160. https://doi.org/10.1016/j.ress.2016.10.033
30. Kong X, Gao L, Ouyang H, Li S (2015) Solving the redundancy allocation problem with multiple strategy choices using a new simplified particle swarm optimization. Reliab Eng Syst Safety 144:147–158. https://doi.org/10.1016/j.ress.2015.07.019

31. Byun JE, Noh HM, Song J (2017) Reliability growth analysis of k-out-of-N systems using matrix-based system reliability method. Reliab Eng Syst Saf 165:410–421. https://doi.org/10.1016/j.ress.2017.05.001
32. Endrenyi J, Andrew GJ (2006) Aging, maintenance and reliability- approaches to preserving equipment health and extending equipment life. Power Energy Maga 4(3):59–67
33. Liu X, Zhang T (2011) Reliability engineering work flow and application in product development process. In: International conference on quality, reliability, risk, maintenance & reliability. IEEE, China, pp 1–4
34. Schoenig R, Aubry JF, Cambois T, Hutinet T (2006) An aggregation method of Markov graphs for the reliability analysis of hybrid systems. Reliab Eng Syst Saf 91(2):137–148
35. Kim MC, Seong PH (2006) An analytic model for situation assessment of nuclear power plant operators based on Bayesian inference. Reliab Eng Syst Saf 91(3):270–282
36. Rajpal PS, Shishodia KS, Sekhon GS (2006) An artificial neural network for modeling reliability, availability and maintainability of a repairable system. Reliab Eng Syst Saf 91(7):809–819
37. Davis P, Burn S, Moglia M, Gould S (2007) A physical probabilistic model to predict failure rates in buried PVC pipelines. Reliab Eng Syst Saf 92(9):1258–1266
38. Zhang YL (2008) A geometrical process repair model for a repairable system with delayed repair. Comput Math with Appl 55(8):1629–1643
39. Haarla L, Pulkkinen U, Koskinen M, Jyrinsalo J (2008) A method for analysing the reliability of a transmission grid. Reliab Eng Syst Saf 93(2):277–287
40. Leicester RH, Wang CH, Cookson LJ (2008) A reliability model for assessing the risk of termite attack on housing in Australia. Reliab Eng Syst Saf 93(3):468–475
41. Galante G, Passannanti G (2009) An exact algorithm for preventive maintenance planning of series-parallel systems. Reliab Eng Syst Saf 94(10):1517–1525
42. Garg S, Singh J, Singh DV (2010) Mathematical modeling and performance analysis of combed yarn production system: Based on few data. Appl Math Model 34(11):3300–3308
43. Kumar N, Tiwari PC, Sachdeva A (2020) Petri Nets modelling and analysis of the Veneer Layup system of plywood manufacturing plant. Int J Eng Model 33(1–2):95–107. https://doi.org/10.31534/engmod.2020.1-2.ri.07v

Shop Floor Productivity Enhancement Using a Modified Lean Manufacturing Approach

Varun Tripathi, Suvandan Saraswat, GirishDutt Gautam, and Dhananjay Singh

Abstract Constantly improvement in shop floor management has become a necessity in today's worldwide industries. Various approaches are used to fulfill this requirement, mainly including lean manufacturing, kaizen, total quality management, six sigma, and lean six sigma approaches. These approaches are used to improve shop floor management and optimization of resources in the present industrial environment. In the present article, the authors developed a modified lean manufacturing approach to enhance the shop floor productivity through optimization. The purpose of this research article is to suggest an approach to improving and optimizing production on the shop floor by research work have done in the area of the lean manufacturing approach. The authors of the current research work strongly believe that the present research method will be more profitable for young researchers and industry people to improve production on the shop floor.

Keywords Lean manufacturing · Shop floor management · Process improvement · Non-value-added activities · Process optimization

1 Introduction

Process improvement is one of the most important needs of industries at a worldwide level. Several techniques are used to accomplish this need. Lean manufacturing (LM), kaizen, total quality management (TQM), six sigma, and lean six sigma (LSS) approaches are including in those techniques. LM is a prevalent technique in shop floor management and is also used for the optimization of resources [1]. As yet, LM was implemented in several industrial sectors such as mechanical, mining, mining

V. Tripathi
Accurate Institute of Management and Technology, Greater Noida, India

S. Saraswat
JSS Academy of Technical Education, Noida, India

G. Gautam (✉) · D. Singh
Mangalmay Institute of Engineering and Technology, Greater Noida, India

A. K. Dubey et al. (eds.), *Recent Trends in Industrial and Production Engineering*, Lecture Notes in Mechanical Engineering,
https://doi.org/10.1007/978-981-16-3135-1_23

machinery, pharmaceutical, defense, and automobile. There are several tools used in lean manufacturing and it depending on the production constraints and customer requirements in terms of product [2]. Here, constraints mean resources availability, delivery time, and production cost. Due to the global competitive environment and a plethora of problems, productivity enhancement has to challenge for the last few decades. Therefore, industries must continually need for an efficient technique to improve shop floor management.

Several researchers have suggested an efficient technique to improve production on the shop floor. Vinodh et al. [3] have suggested that the value stream mapping (VSM) technique was an efficient technique because the reductions have been obtained in production lead time (PLT), total cycle time (TCT), total uptime (TUT), defects, and work in progress (WIP). Rahani et al. [4] have investigated the implementation of the VSM in the automobile component manufacturing industry. The result revealed that hidden non-productive activities on the shop floor like stacking of parts, higher WIP, and more cycle time were eliminated by VSM implementation. Singh et al. [5] have presented VSM as an efficient technique by improvement obtained in an auto-parts manufacturing organization. The result of the study showed a reduction in WIP inventory, PLT, and cycle time. Jasti et al. [6] discussed how improvements could be made in the implementation of a LM approach in the industries. The result of the study proved that lean manufacturing brings out a positive impact on production in form of productivity improvement in terms of customer satisfaction.

Rohani et al. [7] have discussed a study of production line improvement in the chemical industry through VSM. The result of the study revealed that reduction has been obtained in TLT 8.5 days to 6 days and in value-added time (VAT) from 68 to 37 min. Chelbus et al. [8] have identified the three main pillars of a mine and that is the environment of work improvement, planned and autonomous maintenance, and standards in development. As a result of this work, it has been found that lean manufacturing as a technique can be applied to achieve better results in any other mining industry. Garre et al. [9] have identified production problems that occurred during the performance of activities in the production line and eliminated those production problems by the implementation of LM techniques. The result of the study reveals that improvement has been obtained in welding time. Diaz et al. [10] have studied the cycle time of aircraft assembly using LM. The result of the study reveals that reduction has been obtained in cycle time 20% from LM implementation, and 67% from automation implementation. Sharma et al. [11] discussed the implementation of the lean approach on the shop floor of Indian SMEs. As a result of this work, it was found that the implementation of the 5S technique on the SME shop floor effectively improved productivity and also improved the working environment of the employees.

Dadashnejad et al. [12] have investigated the impact of improvements achieved by VSM on overall equipment effectiveness. In the study, a structured questionnaire was designed. The analysis of the questionnaire concluded that VSM can increase performance, machine availability rate, and product quality in the industries surveyed. Gopi et al. [13] have identified wastes in an auto-ancillary unit using VSM. In the study working condition of the auto-ancillary unit has been drawn for the identification and elimination of waste and determines the ideal state of working. The result revealed that manufacturing has been improved by 3.82% in 3 years. Esa et al. [14] discussed how to improve productivity level and production process time in an automotive manufacturing industry. As the result, the reduction has been obtained in setup time for changeover processes from 45 to 28 min as the result of this study. Boateng-Okrah et al. [15] have investigated the level of TQM implementation in the mining industry in Ghana. In this study, a statistical analysis was conducted for knowing the level of TQM in the industry from a questionnaire collection of 60 employees. The analysis of the survey reveals that 42% of employees agreed that TQM is being deployed in the industry. Seth et al. [16] discussed how to achieve lean implementation in heavy-duty power transformer manufacturing environments by implementation of VSM; and Taguchi's technique was also used in this study to improve. Systematic questioning techniques and shop floor observation were used for data collection. The study results showed a 17.3% reduction in cycle time, 29.78% in non-value-added activities, and 8.48% in value-added activities. The literature review on process improvement techniques implementation on shop floor management shows that there were only a few works has been performed on the improvement of production with optimization of resources since the last decade. It has been also found that productivity enhancement cannot be achieved by the implementation of process improvement techniques and process optimization without a strategy on the shop floor [17–20].

2 Research Objective

The main objective of this research work is to explore the effect of applying process improvement techniques in a case example of the industry's loader vehicle manufacturing process. The main steps taken to carry out the present work are enumerated below:

(a) The beginning of the research work is to collect information on the working condition in the industry's loader vehicle manufacturing process through discussion with employees and observing problems in shop floor management.
(b) This step identified the most frequently occurring waste and determines reasons using the fishbone diagram.
(c) Observed how the implementation of process improvement techniques could bring value to the production process.
(d) Which technique would bring improvement in shop floor management and optimization of resources?
(e) Finally, a conclusion is presented considering improvements achieved by the implementation of lean manufacturing technology in the company's production.

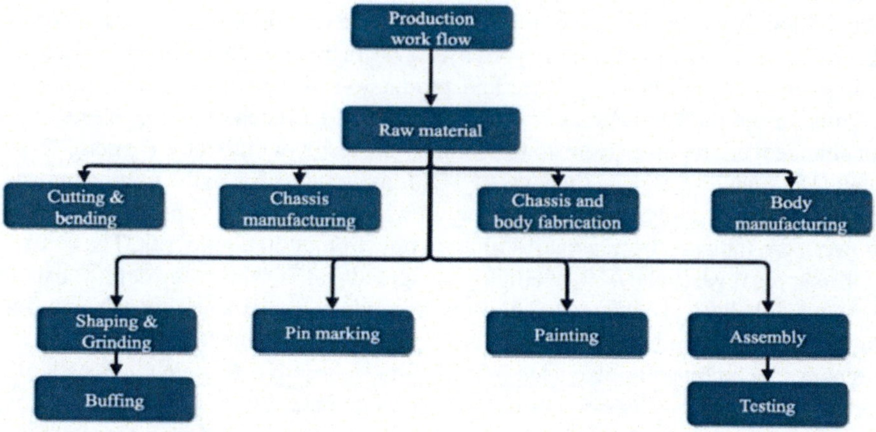

Fig. 1 Flow chart of production processes

3 Production Description

The present work has been carried out in a loader vehicle manufacturing industry located at the National capital region, India. This vehicle is in high demand due to environmental issues, low maintenance costs and less requirement of labor. Figure 1 shows a flow chart of the major production process involved in the manufacturing of the vehicle.

4 Implementation of Process Improvement Technique

LM is a prevalent process improvement technique and it has been mostly implemented in previous research work based on shop floor management [21–24]. The authors studied the literature review to understand the working style of LM, and It was found that LM first identifies the cause of waste generation and then tries to eliminate waste by taking appropriate action [25–27]. At present research work, the loader vehicle manufacturing process was studied and production information has been collected by observation and discussion. Table 1 shows the production condition of the shop floor.

The loader vehicle manufacturing process was categorized into subsections and then focusing on individual work activities. Productivity improvement depends on the elimination of waste and reduction of the processing time in the process [28–30]. The shop floor problem was analyzed by implementing lean techniques. The production problems have been identified using the fishbone diagram. The fishbone

Table 1 Production information on the shop floor

Production data	Quantity
Number of workers	8
Working time	510 min
Number of shifts	2
Production time	560 min
Idle (waste) time	260 min
Number of processes	18
Production per day	5
Break Time	45 min

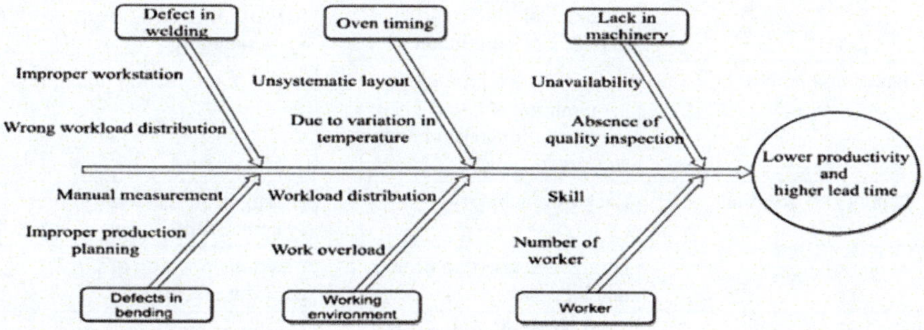

Fig. 2 Fishbone diagram

diagram of waste in the production processes on the shop floor that was obtained from discussing and brainstorming is shown in Fig. 2. Table 2 shows the problems in the production process and the actions taken to eliminate them.

5 Results and Discussion

The results of the present work obtained from a comparative analysis of the working conditions of the shop floor. The result of the study proves that LM is indeed a superior technique than other process improvement technologies. Figure 3 shows improvements obtained in production on the shop floor.

It has been observed that lean manufacturing implementation in the present case study results in increased productivity which was possible by identifying all types of non-value-added activities (waste). Lean manufacturing has also proved to be helpful in identifying waste and suggesting an effective action for their elimination.

Table 2 Suggested action for the elimination of problems

Process	Problem	Action
Cutting and bending	(i) More workstation (ii) Large distance between station (iii) Lack of workers (iv) Lack of machinery	(i) Design a new shop floor layout (ii) Arrange all machinery
Chassis manufacturing	(i) More workstations (ii) Higher changeover time (iii) Lack of setup (iv) Improper location of equipment	(i) They can arrange a new platform of chassis welding (ii) Decided the location of equipment
Body manufacturing	(i) The higher number of defects (ii) Lack of equipment	(i) Improved planning (ii) Arranged equipment systematically
Chassis and body fabrication	(i) Ineffective production planning (ii) Lack of material handling equipment	Improve production planning
Shaping and grinding	Lack of machinery	Arranged all machinery systematically
Pin marking	Unplanned position of the machine	Fixed the position of pin marking machine
Painting	(i) Lack of shop floor area (ii) Undefined timing	Design a new shop floor layout
Buffing	Lack in number of machineries	Arranged machinery in proper planning
Assembly	Ineffective working plan	Improved planning
Testing	Lack of planning	Improved production planning

6 Conclusions

In present research work, lean manufacturing has been implemented to eliminate waste in the vehicle manufacturing industry. The main conclusion obtained by the present work are discussed below:

1. It has been observed that the process improvement technique can provide effective benefits when applied with lean manufacturing for shop floor management. Because it successfully enhanced productivity and optimize resources.
2. The present research work provided a modified lean manufacturing approach that is capable of improving productivity within limited constraints.
3. The Working time, production time, and idle time have been improved 70 min, 80 min, and 80 min respectively by improvement in production planning.
4. Additional improvements were accrued by shop floor utilization and a reduction of inventory resulting in productivity enhancement.

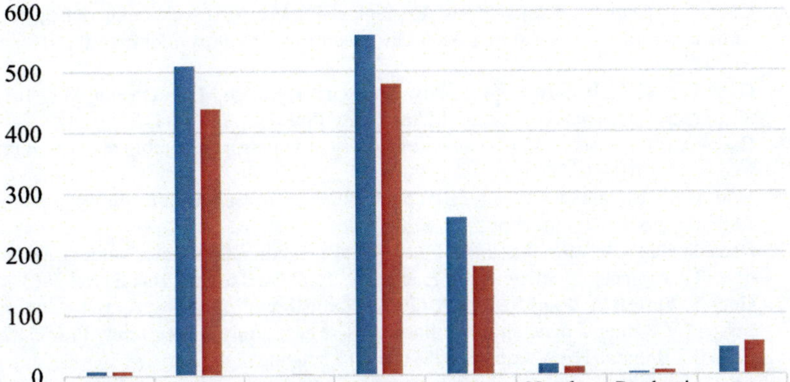

	Number of worker	Working time	Number of shift	Produtio n time	Idle time	Number of process	Producti on per day	Break time
■ Observed data	8	510	2	560	260	18	5	45
■ Improved data	8	440	3	480	180	14	8	55

■ Observed data ■ Improved data

Fig. 3 Production improvement on the shop floor (Time unit—Minute)

5. The authors of the present research work strongly believe that the present study could be beneficial for young researchers and industry person for improvement in shop floor management.

References

1. Ohno T (2014) Toyota production system: beyond large-scale production. CRC Press, London
2. Commission E, Leonardo LLP, Transfer V (2017) Process optimization methods, pp 1–45. https://www.scribd.com/document/366521979/R2-en-COSIMA-Process-Optimization-Methods
3. Vinodh S, Arvind KR, Somanaathan M (2010) Application of value stream mapping in an Indian camshaft manufacturing organisation. J Manuf Technol Manag 21(7):888–900
4. Rahani AR, Al-Ashraf M (2012) Production flow analysis through value stream mapping: a lean manufacturing process case study. Proc Eng 41:1727–1734
5. Singh H, Singh A (2013) Application of lean manufacturing using value stream mapping in an autoparts manufacturing unit. J Adv Manag Res 10(1):72–84. https://doi.org/10.1108/09727981311327776
6. Jasti NVK, Sharma A (2011) Lean manufacturing implementation using value stream mapping as a tool: a case study from auto components industry 5(1)
7. Rohani JM, Zahraee SM (2015) Production line analysis via value stream mapping: a lean manufacturing process of color industry. Proc Manuf 2:6–10. https://doi.org/10.1016/j.promfg.2015.07.002

8. Chlebus E, Helman J, Olejarczyk M, Rosienkiewicz M (2015) A new approach on implementing TPM in a mine—a case study. Arch Civ Mech Eng 15. https://doi.org/10.1016/j.acme.2015.07.002
9. Garre P, Nikhil Bharadwaj VVS, Shiva Shashank P, Harish M, SaiDheeraj M (2017) Applying lean in aerospace manufacturing. Mater Today Proc 4(8):8439–8446
10. Diaz IC, Jin Y, Ares E (2017) Cycle time study of wing spar assembly on aircraft factory. Proc Manuf 13:1019–1025
11. Sharma SS, Shukla DD, Sharma BP (2018) Analysis of lean manufacturing implementation in SMEs: a “5S” technique. Springer, Singapore
12. Dadashnejad AA, Valmohammadi C (2019) Investigating the effect of value stream mapping on overall equipment effectiveness: a case study. Total Qual Manag Bus Excell 30(3–4):466–482
13. Gopi S, Suresh A, John Sathya A (2019) Value stream mapping & manufacturing process design for elements in an auto-ancillary unit—a case study. Mater Today Proc 22:2839–2848
14. Esa MM, Rahman NAA, Jamaludin M (2015) Reducing high setup time in assembly line: a case study of automotive manufacturing company in Malaysia. Proc - Soc Behav Sci 211:215–220
15. Boateng-Okrah E, Appiah Fening F (2012) TQM implementation: a case of a mining company in Ghana. Benchmarking An Int J 19(6):743–759
16. Seth D, Seth N, Dhariwal P (2017) The management of operations application of value stream mapping (VSM) for lean and cycle time reduction in complex production environments? a case study. Prod Plan Control 7287:1–22
17. Tripathi V, Gautam GD., Sarswat S (2020) Process optimization methods for shop floor planning? a study 8(10):244–247
18. Tripathi V, Saraswat S (2018) Lean management implementation in mining equipment manufacturing shop floor. National Conference on Mining Equipment New Technologies Challenges & Applications, Dhanbad (India), pp 7–10
19. Tripathi V, Sarswat S (2018) Lean manufacturing for shop floor of automotive industries: a study. J Exp Appl Mech 9(2):58–65
20. VenkatJayanth B, Prathap P, Sivaraman P, Yogesh S, Madhu S (2020) Implementation of lean manufacturing in electronics industry. Mater Today Proc. https://doi.org/10.1016/j.matpr.2020.02.718
21. Shou W, Wang J, Wu P, Wang X (2020) Lean management framework for improving maintenance operation: development and application in the oil and gas industry. Prod Plan Control. https://doi.org/10.1080/09537287.2020.1744762
22. Sivaraman P, Nithyanandhan T, Lakshminarasimhan S, Manikandan S, Saifudheen M (2020) Productivity enhancement in engine assembly using lean tools and techniques. Mater Today Proc. https://doi.org/10.1016/j.matpr.2020.04.010
23. Amrani A, Ducq Y (2020) The management of operations lean practices implementation in aerospace based on sector characteristics: methodology and case study methodology and case study. Prod Plan Control. https://doi.org/10.1080/09537287.2019.1706197
24. Sutharsan SM, Mohan Prasad M, Vijay S (2020) Productivity enhancement and waste management through lean philosophy in Indian manufacturing industry. Mater Today Proc. https://doi.org/10.1016/j.matpr.2020.02.976
25. Mohan Prasad M, Dhiyaneswari JM, Ridzwanul Jamaan J, Mythreyan S, Sutharsan SM (2020) A framework for lean manufacturing implementation in Indian textile industry. Mater Today Proc. https://doi.org/10.1016/j.matpr.2020.02.979
26. Balamurugan R, Kirubagharan R, Ramesh C (2020) Implementation of lean tools and techniques in a connecting rod manufacturing industry. Mater Today Proc. https://doi.org/10.1016/j.matpr.2020.03.702
27. Gopi S, Suresh A, John Sathya A (2019) Value stream mapping & manufacturing process design for elements in an auto-ancillary unit-A case study. Mater Today Proc 22:2839–2848
28. Masuti PM, Dabade UA (2019) Lean manufacturing implementation using value stream mapping at excavator manufacturing company. Mater Today Proc 19:606–610. https://doi.org/10.1016/j.matpr.2019.07.740

29. Mundra N, Mishra RP (2020) Impediments to lean six sigma and agile implementation? An interpretive structural modeling. Mater Today Proc. https://doi.org/10.1016/j.matpr.2020.04.141
30. Suhardi B, Anisa N, Laksono PW (2019) Minimizing waste using lean manufacturing and ECRS principle in Indonesian furniture industry. Cogent Eng 6(1):1–13

Simulation Model for Productivity, Production and Cost Improvement in an SME

P. M. Kinge and U. C. Jha

Abstract Every industry endeavoring its best to turn into smart intellective manufacturing to sustain the competition predicated on development following industry 4.0. Smart intellective manufacturing is predicated on smart intellective decision making. Smart intellective decisions can be taken with lots of paperwork, brainstorming, presentations, etc., and the utilization of these traditional techniques results in decremented productivity. Simulation software is the implement that fortifies astute decision making to amend productivity, production, and cost in a manufacturing unit. The perspicacious decision making is predicated on the analysis of data amassed from various activities of the manufacturing unit. This paper aims to spread vigilance about the utilization of such implements in Indian small medium enterprises (SMEs) for enhancing organizational performance. In this paper, the research work is predicated on modeling the problem of the plastics processing industry utilizing Arena simulation software to amend productivity, production, and cost. The selected manufacturing unit produces around 30 components and supplying them mainly to the furniture-making industry. The conception is to develop a simulation model that can be facilely utilized by the working executives of the SME. By changing various process parameters, different scenarios can be obtained with different outputs. The decision to select the best scenario with the optimum output can be utilized for implementation on the shop floor.

Keywords SMEs · Arena · Simulation model (SM) · Framework

1 Introduction

For so many years, various advanced implements are in industrial practices for system modeling. They are efficaciously used to model various industrial quandaries to ascertain various scenarios with transmuting parameters. All the scenarios are compared to get the best one afore being implemented in authentic practice. Arena simulation

P. M. Kinge (✉) · U. C. Jha
Lovely Professional University, Phagwara, Punjab, India

A. K. Dubey et al. (eds.), *Recent Trends in Industrial and Production Engineering*, Lecture Notes in Mechanical Engineering,
https://doi.org/10.1007/978-981-16-3135-1_24

is utilized to model the effect of the number of machines on resource utilization and the solutions obtained are acclimated to frame the dispatch rules Yang and Hsieh [1]. The majority of manufacturing units are endeavoring their best to become keenly intellective and want to follow the framework proposed by various researchers following Industry 4.0. Considerable research is done for developing a framework by many researchers concerning the practices followed in the industries. One such framework, Operator–Workstation Interaction 4.0 is developed by Golan et al. [2]. This framework consists of three subsystems: 1. observing and accumulating data on working of operator and workstation, 2. analyzing the data for engendering output, 3. replication system. As a component developing models of manufacturing systems, the discrete event simulation model is developed with all multifarious features which in turn can develop various scenarios afore being implemented. With the affordability of the simulation software, their utilization plays a vital role in the amelioration of efficiencies across various verticals in the SMEs Aggogeri et al. [3].

Recently, since last decade utilization of simulation software got momentum in many industries to develop the different strategies by varying input parameters. There is an ease in the utilization of this software as compared to the mathematical models developed for various industrial problems.

A generalized production inventory mathematical model is developed by Yang [4]. The objective of this model is to find optimum production restart and stop time by keeping the cost as low as possible. The approach utilized in developing such a mathematical model differentiates considerably with the approach by the researcher or industrial utilizer who develops models with the simulation software. Similarly, a mathematical model was developed by Zhang et al. [5] and explicates the integration of the production inventory system. The objective of this model is to ameliorate supply chain efficiency. When you compare the development of this model with the model which can be developed utilizing simulation software with the same problem, you may find that the second approach is very efficient and time-preserving over the first one.

Simulation implements are widely utilized in developing models in the product design process, for minimizing development time, truncate errors, and optimize the process. Such an implement is utilized by Tosello et al. [6] for the optimization of raw material input into the hopper of the powder injection molding process. A logical framework can additionally be developed to ascertain the solution to various industrial problems. One such framework is developed by Huang et al. [7] and predicated on the reproductive property of probability distribution to ascertain the optimum production rate and lot size.

Lie and Jia [8] developed EPQ models to ascertain the optimum production lot size and backorder quantity. They utilized a traditional approach to develop the EPQ mathematical model and tested two numerical with authentic industrial data. Wang et al. [9] developed a framework for task scheduling. The framework integrates all the smart resources of the manufacturing unit so that all the activities are coordinated in authentic time. The utilization of advanced technology and their interface with each other is the base of this framework. The framework proposed is tested with a case study.

The role of SMEs in the economic development of any country and adaptation of Industry 4.0 by the SMEs and its impact on their prosperity along with risk associated is well explicated by Moeuf et al. [10]. The effect of Industry 4.0 technologies (I4T) on sustainable organizational performance (SOP) with lean manufacturing practices (LMP) as the variable is a component of the research done by Kamble et al. [11].

Alkhoraif et al. [12] elongated the Systematic Review Methodology (SRM) for understanding the effect of lean implementation in SMEs. A mathematical model was developed by Giri and Sharma [13] for determining optimum production and shipment policy. Li [14] developed a mathematical model of production inventory to ascertain optimum inventory levels and production rates to minimize total cost.

2 Problem Statement

The SME under consideration for research work is the plastic processing industry producing around 30 components and supplying mainly to the furniture-making industry. The unit is mostly flooded with orders containing varying demands for sundry products produced by the industry. For each product, there is discrete mold and it requires to be set on a particular machine. There are a total of 13 machines with two types of capacities 100-ton and 50-ton. A total of 3 machines are with 100-ton capacity and 10 machines with a 50-ton capacity. Lots of paperwork needs to be done by the production executive for deciding which machine to stop for production and when to stop as the number of products is more than the number of machines. The product demands are varying with time. The simulation model abbreviates this paperwork to a great extent additionally helps in is deciding the optimum quantity for production so that all the customer demands are met. There is cost associated with the inventory of so many products; the simulation model answers the optimum inventory for each product.

3 Action Plan for Development of the Simulation Model

The simulation model (SM) is divided into two components, inventory, and production. The inventory part of the SM accepts customer orders with product demands as input, checks the demands against inventory for that product, and sets the product targets accordingly. The product demand and rejection data are stored in the excel file during the simulation run. This data is utilized to find rejection for each product. The production part of the SM contains an individual model for 13 machines. There are only two types of machines 100-ton and 50-ton. Total 3 machines are 100-ton and the remaining 10 are 50-ton plastic injection molding. The products are produced on these 13 machines according to their production targets. Each machine's production data is stored in the excel file during the simulation run. The data is utilized for deciding the loading and unloading of molds on the 13 machines. The flow chart for the development of simulation model is shown in Fig. 1.

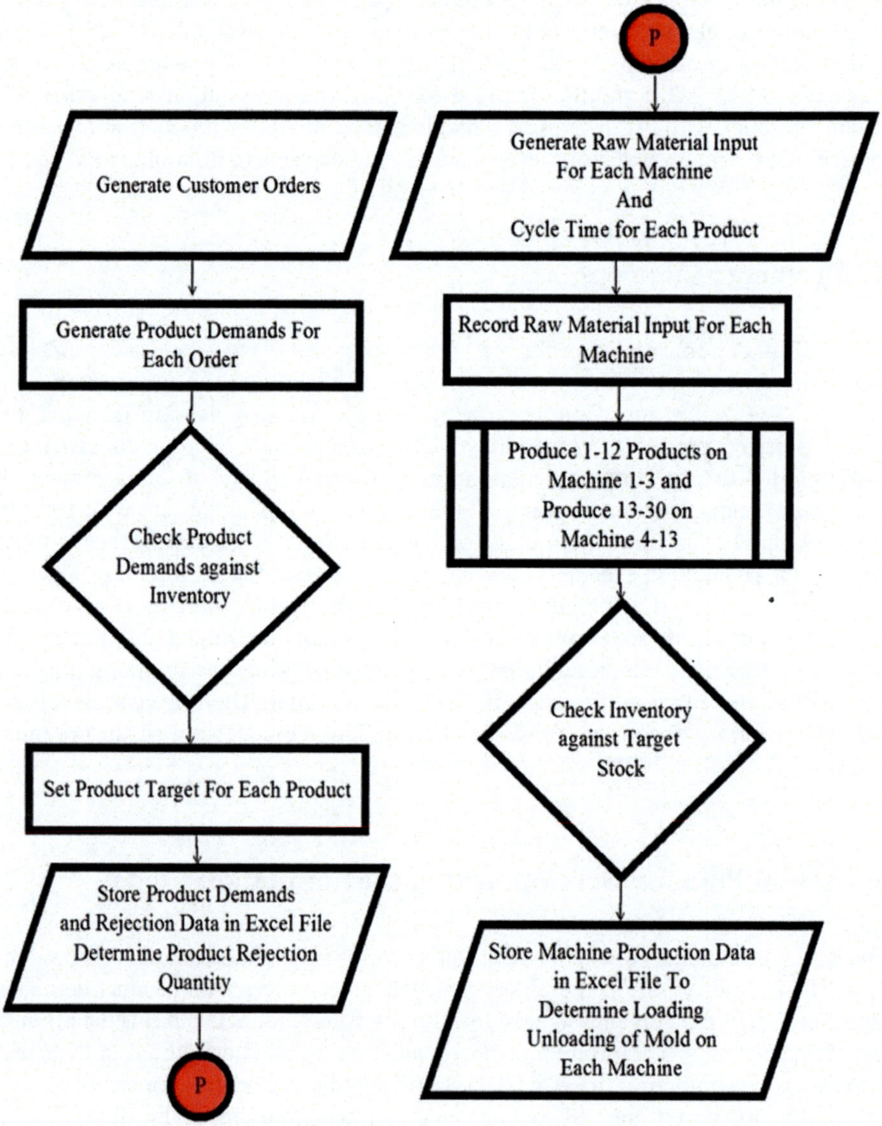

Fig. 1 Action plan for development of simulation model

4 Design of the Simulation Model

The simulation was designed utilizing Arena simulation software. For the inventory part of the model, the input is inter-arrival times for customer orders, product demands for each order. Similarly, for the production part of the model, the input is the inter-arrival time for raw material on each machine and cycle time for each product. The factory under consideration runs in 3 shifts of 8 h each for 6 days in a week. The simulation run is taken for 30 days. The statistical distribution expression for each input is developed based on data accumulated from the plastics processing industry under consideration. A total of 13 machines are there and 30 products are produced on these machines. 12 products are produced on 3 machines with 100-ton designation, and the remaining 18 products are produced on 10 machines with the 50-ton designation. Without this simulation model, the production management of 30 products on 13 machines is tedious. The simulation model provides a solution for time-based loading and unloading of product mold on a particular machine so that the rejection of the orders is minimized.

5 Results and Discussion

The data recorded as a component of the output in the excel file is for product demands and rejection. An excel file was created with a macro written for finding the rejection for each product for a simulation run of 30 days. Concerning the data product numbers 2, 3, and 20 got maximum rejection. The simulation model has built an implement denominated process analyzer. By utilizing this implement, one can vary the input controls and after run replications are engendered in the form of output. Here, in our case, it is required to reduce the rejection of product numbers 2, 3, and 20. The following self-explanatory Table 1 explains the input control and output responses for the 3 scenarios run in the process analyzer of Arena simulation. As product numbers, 1 to 12 are produced on machines 1 to 3, and product numbers 13 to 30 are produced on machine 4 to 13. With the process Analyzer, it is very easy to find optimum inventory and target stock for each product by running the number of scenarios within a few minutes.

The production data of each machine is recorded in an excel file. A time-based decision can be taken for loading and unloading of mold on the categorical machine. As the simulation model is run for 30 days with new customer order entering into the system with every 6 days. The production on 13 machines for 30 products took place for 615 h. The output generated by the simulation model is validated by implementing the solution for 2 consecutive months in the SME under consideration. Table 2 shows the production of product numbers 1 to 12 on 3 machines (100 tons each).

Table 3 shows production of product number 13 to 30 on remaining 10 machines (50 ton each).

Table 1 Product target and product rejection scenario

	Input controls						Output response		
	Inventory(2)	Target stock (2)	Inventory (3)	Target stock (3)	Inventory (20)	Target stock (20)	Products rejected Qty (2,2)	Products rejected Qty (3,2)	Products rejected Qty (20,2)
Scenario 1	600	5000	600	3500	1122	2200	3149	4597	4172
Scenario2	1000	7000	1000	5000	1800	4000	3149	2414	1375
Scenario3	600	4500	600	4500	1122	4000	3149	2414	1375

Table 2 Production on 3 machines (100 ton each)

Machine 1			Machine 2			Machine 3		
Product No	Qty Produced	Time	Product No	Qty Produced	Time	Product No	Qty Produced	Time
1	2800	3.26	2	3800	3.92	3	2300	2.78
6	5332	6.55	7	7527	7.33	5	5346	6.09
1	2030	146.37	6	3142	145.97	3	2068	146.49
4	299	290.64	12	995	147.80	2	3844	291.97
6	3008	292.50	1	2886	291.35	10	2471	296.54
2	447	432.47	7	2086	292.30	4	46	432.42
7	7541	435.89	12	1040	294.22	6	4922	435.47
5	568	576.37	5	3618	434.26	7	4055	577.86
10	1014	578.25	12	1383	436.82	12	458	578.71
			4	381	579.31			

Note Time is in hours

Table 3 Production on 10 machines (50 ton each)

Machine 4			Machine 5			Machine 6		
Product No.	Qty produced	Time	Product No.	Qty produced	Time	Product No.	Qty produced	Time
15	3300	2.59	16	3504	2.76	18	2024	1.44
26	1154	433.21	29	242	433.27	18	2895	578.05
28	1634	579.96	25	398	578.68			
Machine 7			**Machine 8**			**Machine 9**		
Product No.	**Qty produced**	**Time**	**Product No.**	**Qty produced**	**Time**	**Product No.**	**Qty produced**	**Time**
18	2823	146.00	20	1924	147.18	17	893	290.98
15	2717	290.14	16	2649	290.10	20	1853	579.07
15	1865	433.48	17	42	432.15			
16	1620	577.29	17	639	578.11			
Machine 10			**Machine 11**			**Machine 12**		
Product No.	**Qty produced**	**Time**	**Product No.**	**Qty produced**	**Time**	**Product No.**	**Qty produced**	**Time**
18	2204	289.55	26	742	288.78	18	695	432.50
16	512	432.41	14	639	434.36	14	6	576.04
26	1766	577.83	27	360	578.62	29	314	577.65
Machine 13								
Produet No.	**Qty produced**	**Time**						
25	161	497.50						
15	823	614.58						

The objective of batch production is to produce an optimum quantity of each product at the right time so that the demand for maximum products is met. The output obtained in the excel file for production facilitates the same.

6 Conclusion and Future Scope

The productivity of the shop floor manager has been improved to a great extent by eliminating the tedious job of manual production scheduling. There is an improvement on the production front, the optimum inventory, and target stock for each product can be found by running various scenarios with inventory and target stock as input controls to produce the replications in the form of rejection quantity for each product. Last but not the least there is cost associated with the inventory of the products, and when you hold an optimum quantity of inventory for each product obviously the total inventory cost is optimum.

As a part of quantitative findings, the simulation model assists to process the huge machine data in few minutes, in our case it is for 13 machines, in other production floor it may consist of around 100 machines. The machines may run virtually for few days to years in the simulation model and generate the required output. As a part of qualitative findings, the productivity of the shop floor manager improved to great extent by eliminating the preparation of manual production schedule. Since there is change from manual work to the solution obtained from simulation model, so there are no comparison results between the existing and proposed product.

This simulation model contains two components inventory and production. As in normal practice, there is customary machine maintenance or breakdown, so there is scope for integrating a maintenance model into this existing model. The reports produced by this simulation model provide insights into resource utilization. The cost analysis can easily be obtained in these reports if the cost input data is provided by the industry. There is scope affixing the operator schedule to each machine and ascertain their instantaneous utilization. The effect of addition or reduction of resources like human, machine, material, etc., on the cost can easily found and optimum numbers can be found by running various scenarios.

References

1. Yang C, Hsieh C (2014) A production scheduling simulation model for improving production efficiency A production scheduling simulation model for improving production efficiency. Cogent Eng 31(1):1–8
2. Golan M, Cohen Y, Singer G (2019) A framework for operator—workstation interaction in Industry 4.0 7543:0–12
3. Aggogeri F, Faglia R, Mazzola M, Merlo A (2015) Automating the simulation of SME processes through a discrete event parametric model regular paper
4. Yang H (2011) A partial backlogging production-inventory lot-size model for deteriorating items with time-varying production and demand rate over a finite time horizon 42(8):1397–1407
5. Zhang Q, Luo J, Duan Y (2014) Buyer– – vendor coordination for fixed lifetime product with quantity discount under finite production rate. Int J Syst Sci, 37–41
6. Tosello G, Marhöfer DM, Islam A, Müller T, Plewa K, Piotter V (2019) Comprehensive characterization and material modeling for ceramic injection molding simulation performance validations
7. Huang Y, Wang R, Ho J (2015) Determination of optimal lot size and production rate for multi-production channels with limited capacity. Int J Syst Sci 37–41
8. Liao G-L (2013) Optimal economic production quantity policy for randomly failing process with minimal repair, backorder and preventive maintenance. Int J Syst Sci 37–41
9. Wang X, Yew A, Ong S, Nee A (2019) Enhancing smart shop floor management with ubiquitous augmented reality. Int J Prod Res 1–16
10. Moeuf A, Lamouri S, Pellerin R, Giraldo S, Valencia E, Eburdy R (2019) Identification of critical success factors , risks and opportunities of Industry 4.0 in SMEs. Int J Prod Res 1–17
11. Kamble S, Gunasekaran A, Dhone NC (2019) Industry 4.0 and lean manufacturing practices for sustainable organisational performance in Indian manufacturing companies. Int J Prod Res 1–19
12. Alkhoraif A, Rashid H, Mclaughlin P (2019) Lean implementation in small and medium enterprises: literature review. Oper Res Perspect 6:100089

13. Giri BC, Sharma S (2014) Lot sizing and unequal-sized shipment policy for an integrated production-inventory system. Int J Syst Sci 37–41
14. Li S (2014) Optimal control of the production – inventory system with deteriorating items and tradable emission permits. Int J Syst Sci 37–41

STL Generation from Point Cloud Data with User-Controlled Triangulation for Additive Manufacturing

Sourabh Chasta, Narendra Kumar, Vishal Francis, and Prashant K. Jain

Abstract In the last two decades, reverse engineering (RE) and additive manufacturing (AM) have been emerged together as a tool to fabricate intricate parts in minimum time. RE captures the point cloud data and converts it to standard triangulation language (STL) which is compatible to AM systems. The existing approaches to generate STL either use costly software or use decade-old algorithms, which are ambiguous and difficult to understand. The current research proposes a mesh construction algorithm (MCA), which endeavors to provide a detailed, simplified and modified algorithm to generate STL from unorganized point cloud data using Delaunay triangulation and its properties. An additional feature of proposed MCA is that user can control shape of mesh, and chooses best STL according to the requirement. Moreover, this article describes a descriptive algorithm to create STL from cloud data in easy way using MATLAB and its function. The generated STL files are compatible with other software and parts can be 3D printed directly using these files.

Keywords Additive manufacturing · Reverse engineering · CAD · Mesh construction algorithm

S. Chasta · P. K. Jain
Mechanical Engineering Discipline, PDPM Indian Institute of Information Technology, Design and Manufacturing Jabalpur, Jabalpur, India

N. Kumar (✉)
Industrial and Production Engineering Department, Dr. B R Ambedkar National Institute of Technology, Jalandhar, India
e-mail: kumarn@nitj.ac.in

V. Francis
School of Mechanical Engineering, Lovely Professional University, Jalandhar, India

A. K. Dubey et al. (eds.), *Recent Trends in Industrial and Production Engineering*, Lecture Notes in Mechanical Engineering,
https://doi.org/10.1007/978-981-16-3135-1_25

1 Introduction

Additive manufacturing (AM) is an emerging domain of engineering, which has gained popularity due to ability of fabricating complex objects directly from CAD model. AM provides freedom to affordable manufacturing of a complex shape in less time. It reduces the lead time of product development cycle by providing a prototype prior to start actual manufacturing. Before coming into hands, the produced prototype from AM has to pass through various steps in a chained system. Generation of CAD model is the primary step and can be performed through various software. Sometimes, the available software for modeling shows incompetency in creating CAD models of the intricate shapes like architecture buildings, unsymmetrical artifacts, human bones, etc. The modeling of these shapes in dedicated software may be time consuming and uneconomical. In this situation, reverse engineering (RE) technique is utilized to generate the CAD models of unsymmetrical and complex geometries. It collects physical data information of an object via scanning operations using contact or non-contact type machines. The coordinate measuring machine is commonly used in contact type scanning machine, whereas laser scanners come under the non-contact type scanning machines. These measuring machines provide coordinates of an object, which is commonly referred as point cloud data. The point cloud data is further processed and converted in a supportive mesh file for additive manufacturing.

The mesh file format should support both modeling and printing software or it should be written in a universally accepted format. Standard triangulation language (STL) file format is used as the de facto standard for mesh files. It stores only surface information of the 3D object described as an unstructured triangulated surface. The conversion of point cloud into STL is a complex and difficult task, but there are some commercial software like Geomagic Control, point cloud processing of 3DReshaper, etc., available in the market to accomplish this purpose. Moreover, there are some software freely available but they are less efficient in generating defect free STL. However, sometimes there are compatibility issues occurred during the STL exchange with other software. Thus, more research needs to be carried out for developing an economical and compatible platform to generate the error free STL directly from the point cloud.

Thus, the present paper aims to develop an algorithm to generate the STL file via proposed mesh construction algorithm (MCA) with user-controlled tessellation. The MCA can be used reverse engineering, virtual reality and computer vision, moreover, it can be used to integrate with additive manufacturing. The paper also provides a methodology to imply MCA in MATLAB software, which is mostly used by researchers for its advanced technical computing and user-friendly interface. The generated STL files are fabricated using 3D printer in order to validate the present work.

2 Literature Review

Construction of surface for 3D point cloud data was considered as an impossible task in past decades. The fortune, sweep line and Delaunay algorithms were confined to solve only 2D mesh problems. In the recent years, some studies proved the possibilities of generation of mesh for 3D data. Choi et al. [1] presented an optimal triangulation algorithm for 3D discrete points with smoothness surface criterion using Delaunay triangulation (DT). Bernhard et al. [2] proposed a solution to create 3D surface using DT for planar surface and create tetrahedron between two consecutive planar surfaces. They reconstructed the surface of bone using scanned images. Varady et al. [3] identified the RE in industrial applications and discussed about the problems in data acquisition technique and surface construction from wrong data points. Piegl et al. [4] presented a method to reconstruct the model from point data using B-spline curves by approximate the data and controlling the tolerances of knot vectors. Recently, Woo et al. [5] presented a novel method to segment the scanned data by using octree-based 3D grid technique and even works on unorganized data points. After successful creation of 3D surface from scanned data, researchers utilized the potential of RE in additive manufacturing by converting scanned data directly into STL. Chen et al. [6] presented an approach to generate STL file from coordinate measuring machine (CMM) with user control point reduction. Their study was mainly focused on data reduction of point cloud and was limited to create STL for organized data points. Lee et al. [7] developed an algorithm to generate STL by dividing scanned data in smooth, rough and plane regions. Nevertheless, the study showed the generation of STL file using Christiansen algorithm [8] and focused more on refining the prepared STL using data reduction. Recently, more advanced work has been attempted by researchers in which one more step has been reduced in RE technique by directly slicing the point cloud to 3D print the part. Xu et al. [9] presented an improved virtual edge approach to slice point cloud data and form contours. Their focus was on improving the quality of contours by filling the gaps in point cloud by providing additional points. Piegl et al. [10] presented a work on the direct slicing of point cloud. In their approach, slicing and printing were done simultaneously for one layer then another set of data points were considered to prepare next slice. Yuan et al. [11] presented the direct slicing approach to print part contours directly from point cloud without preparing any CAD model or STL. The printing of complex geometries from point cloud using FDM printers was performed. Their approach of part printing was not versatile in nature as it failed to create toolpath for area inside the contours. PLSP-based approach has also been used to develop contour information from point cloud for additive manufacturing. The method proposed is robust as it does not involve the nonlinear optimization [12].

Aforementioned literature shows that various approaches have been presented to convert scanned data into 3D surface or STL. The presented algorithms that generate STL are complex in nature and outdated to process large number of data points. Therefore, present paper aims to develop a modified algorithm of STL generation from any kind of 3D scanned data, either organized or unorganized.

3 Terminologies

- Point Cloud: Group of points situated anywhere in three-dimensional space referred as point cloud. The point clouds refer to the surface points obtained from 3D scanners and coordinate measuring machines.
- Circumsphere and circumcircle: A sphere that touches all vertices of polyhedron and is bound in its shape known as circumsphere. Similarly, circumcircle is used for planar case. Commonly, both terms are used in DT in context of tetrahedron and triangles meshing. The radius of circumspheres or circumcircles is known as circumradius.
- Alpha Shape function: It is an in-built function in MATLAB used to generate triangulated mesh over point cloud data. It uses DT to tessellate the point cloud. It uses alpha radius to construct mesh, which is equivalent to circumradius.
- Boundary Facet function: It is also an in-built function used to generate triangular connectivity list for the given 3D point cloud.

4 Methodology

In order to generate STL, a certain set of points is required. Therefore, in the current study, the point cloud data acquired from various resources are stored in the form of matrix. In this matrix, rows represent the number of points, whereas corresponding coordinates are represented in the columns. Based on this matrix, the mesh construction algorithm (MCA) distinguishes the nature of point cloud and then proceeds further to form a mesh. A well-known Delaunay triangulation (DT) property has been used to construct a mesh over data. Subsequently, STL file has been prepared with the help of the constructed mesh. The aforementioned method of preparing STL is implemented in MATLAB software. The prepared STL is then analyzed through exporting in Solidworks software. After ensuring the quality of STL, final validation has been done using 3D printing software and printer. Physical prototypes are printed using prepared STL to demonstrate the robustness of developed algorithm. The methodology adopted in the current research work is shown in Fig. 1.

Delaunay triangulation is used to tessellate the discrete point in a three-dimensional plane. DT creates tetrahedron for 3D discrete points and triangles for planar points. All tetrahedrons and triangles must follow the Delaunay's property that no other points should fall inside the circumscribed sphere or circle. For 3D, DT creates tetrahedron using 4 discrete point circumscribed in sphere. General equation of sphere in 3D plane is $(x-a)^2+(y-b)^2+(z-c)^2=r^2$, where (a, b, c) is center of sphere and (x, y, z) is any point on the surface of sphere. The unknown parameters used in this equation can be determined using the following steps:

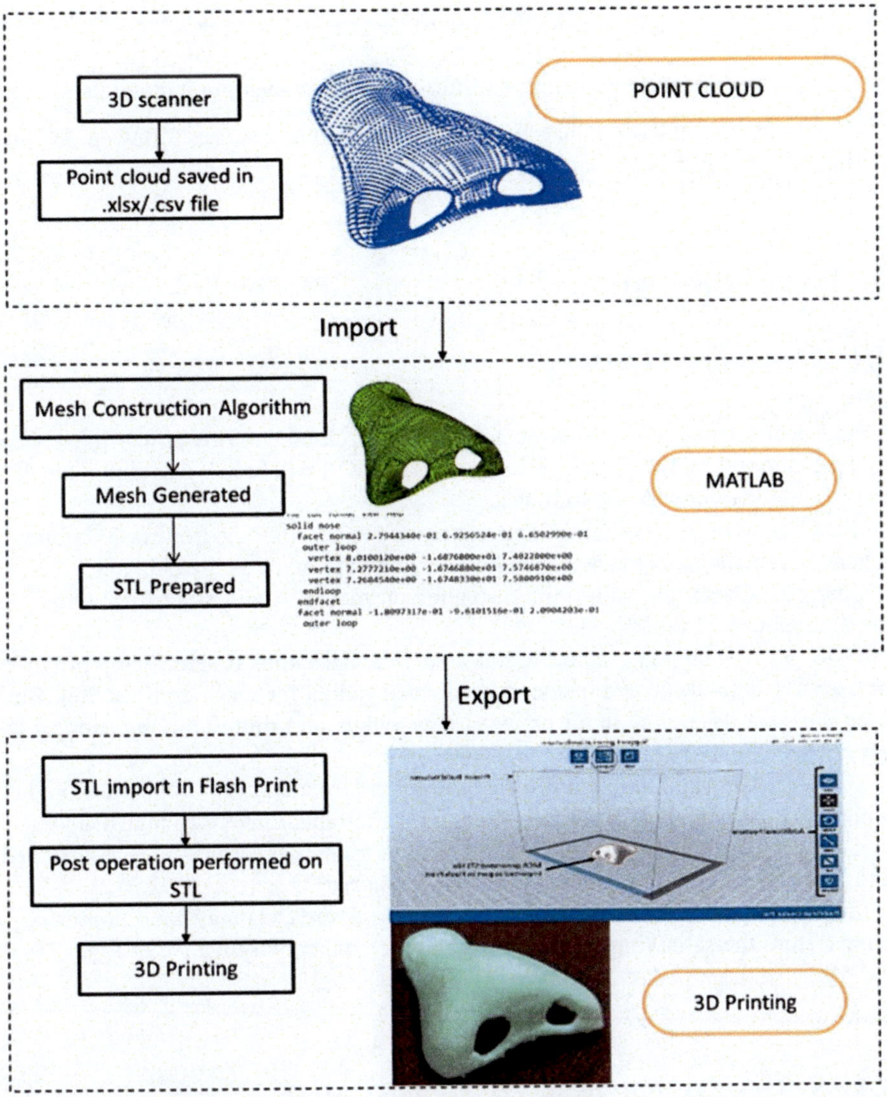

Fig. 1 Methodology

- Select any four points $\{P1, P2, P3, P4\}$ from cloud data and create sphere using above equation. When sphere radius and center are known, check the property of Delaunay on it.
- Check the Delaunay property on circumsphere by satisfying following condition then only tetrahedron will be considered valid.
  $\forall \{x, y, z\} - \{P1, P2, P3, P4\} \exists$ sphere with centre (a, b, c)
  and radius r supports

$$(x - a)^2 + (y - b)^2 + (z - c)^2 > r^2$$

- The joining of valid tetrahedrons results in a surface over point cloud data.

However, for 2D or planar case, same procedure can be followed by just eliminating the given constant dimension.

5 Implementation

5.1 Developed Algorithm

Firstly, decision has to be made whether given point cloud is of two-dimensional body or three-dimensional structure, and accordingly, MCA will be selected automatically. For 3D, algorithm works as follows:

Step (1). Matrix A contains 3D point passed to DT (tetrahedron for 3D points), where circumsphere is made out of 4 points and circumradius is calculated.

Step (2). Check the valid circumspheres on matrix A and store their radius in descending order in matrix B.

Step (3). Decision is to make at this step, whether radius is selected by program or user. By default, program selects median of radius for mesh construction. But, user can give the radius input between max-radius and min-radius by user's past experience with code (Fig. 3).

Step (4). The valid tetrahedrons that satisfy the Delaunay condition are stored in a matrix C in the form of triangular faces, i.e., 1 tetrahedron = 4 triangular faces.

Step (5). By eliminating the common faces, a triangular surface mesh is obtained and their point to point connectivity stored in Matrix D.

Similar steps will be followed for 2D, just check the Delaunay condition using 3 points, store the satisfying points indexes in the connectivity list (refer Figs. 2 and 3).

5.2 STL Generation from Triangulated Mesh

STL is a universally accepted file format in AM for fabricating three-dimensional parts. It has become a "defacto" standard for transferring CAD data into AM software environment. Any system or software related to AM can accept STL as an input file. ASCII and Binary are two formats of STL in which a CAD data can be exported as per user requirement. In the current study, developed triangulated mesh is exported in ASCII file format. The ASCII format is selected as it is human readable. In this format, information about triangle is represented with the help of three coordinates and associated surface normal.

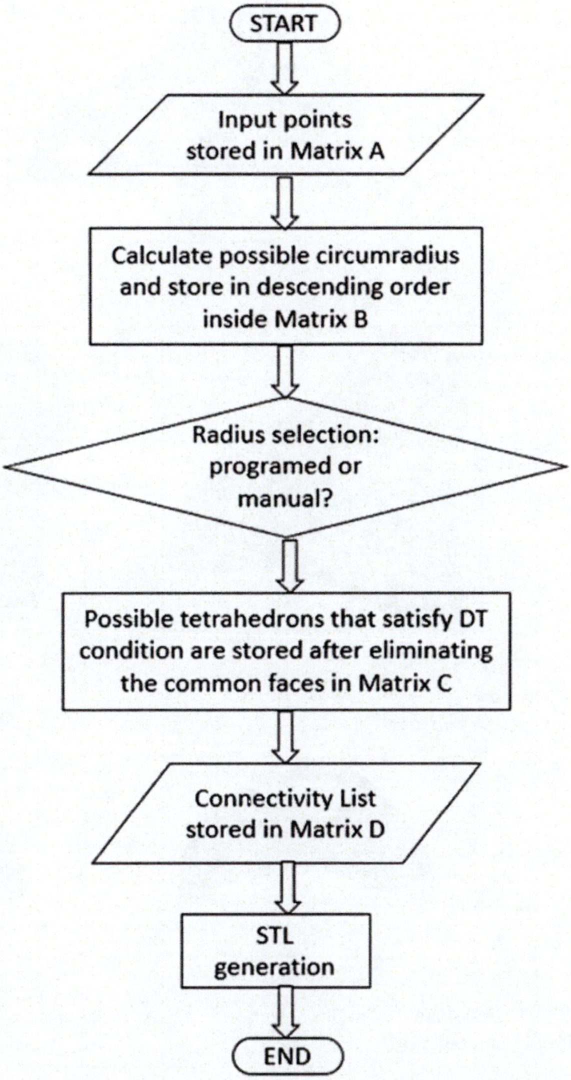

Fig. 2 Steps followed in developed algorithm

In first step, the output obtained from MCA, which contains connectivity list of triangular mesh, is processed to generate STL file. Let n be the number of faces created by MCA, and thus, a loop is started from $i = 1$ to n to make ASCII type STL. On termination of loop, the file is saved with .stl extension. The generated STL is then fed into the Solidworks to observe the quality (Fig. 5). An example of nose was taken and converted into STL file from obtained triangulated mesh, as shown in Fig. 4.

The discussed algorithm can be coded in any language, but MATLAB (2015 or above) provides "Alpha Shape" and "Boundary Facets" functions that ease to

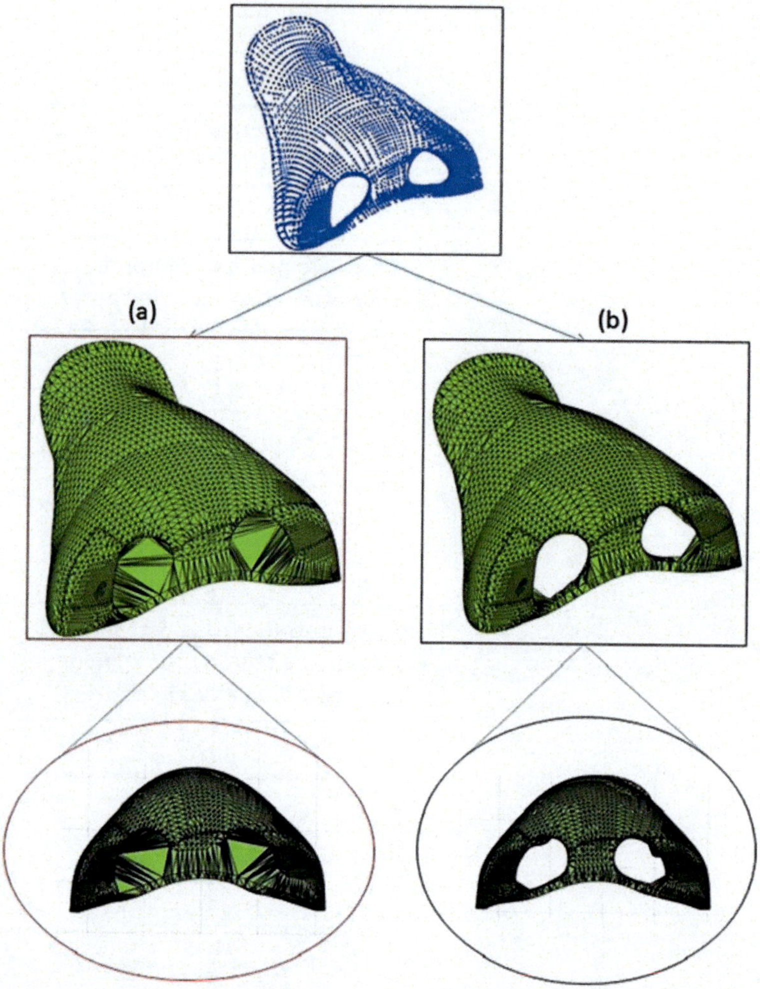

Fig. 3 Importance of radius selection in MCA: mesh created by **a** non-optimum radius, whereas **b** optimum radius

deploy the MCA. Alpha Shape creates tessellation for both 3D and 2D point cloud data. Whereas, Boundary Facets give connectivity list of points. Figure 6 shows the graphical user interface (GUI) of implementation of MCA in MATLAB. Using the developed GUI, user can interact with the point cloud data with ease and can create mesh as per the requirement.

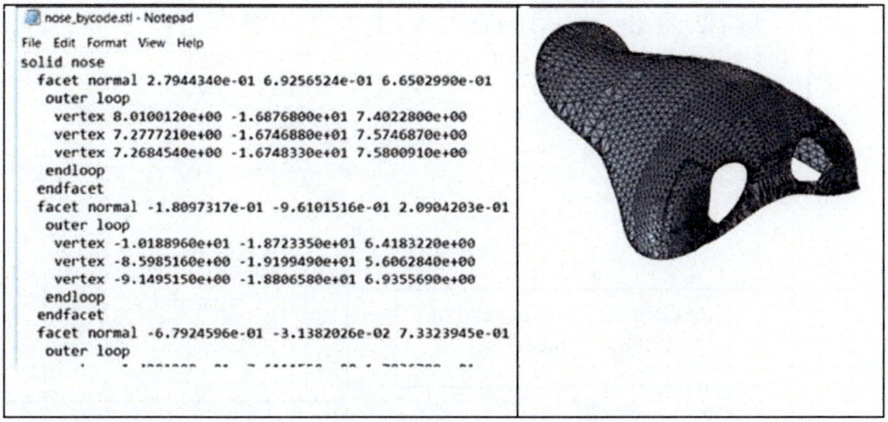

Fig. 4 Shows the generated STL file form MCA opened in (Left) Notepad, (Right) Solidworks

6 Validation

In order to show the robustness of proposed algorithm, STL file generated through MCA has been printed through fused filament fabrication (FFF) AM system. FFF offers part fabrication at relatively low cost as compared to other AM systems. In the current study, a FFF printer FlashForge creator pro along with flashprint software was used to fabricate samples. Figure 7 shows the STL file of nose exported in environment of flashprint software.

Proposed algorithm is able to generate STL for complex structures, which are validated by fabricating through 3D printing. Table 1 represents the fabricated samples and the used point cloud data. Moreover, the processing time is also presented which is measured from exporting point cloud data in MATLAB to STL generation, excluding the generating printing code and manufacturing time. Sample printing was done with acrylonitrile butadiene styrene (ABS) material by considering standard process parameters.

From the results, it can be seen that printed parts have similar structure and shape as possess in point cloud data. It means prepared STL are printable and can be printed by using any standard AM system. Overall, 3D printing results show the robustness of proposed algorithm.

7 Conclusion

In this paper, a generalized methodology of converting point cloud data into STL was presented. mesh construction algorithm (MCA) proposed solution was efficient and able to work on unorganized point cloud of complex shapes. The STL files related to architectural and biomedical sectors were successfully developed and fabricated. The

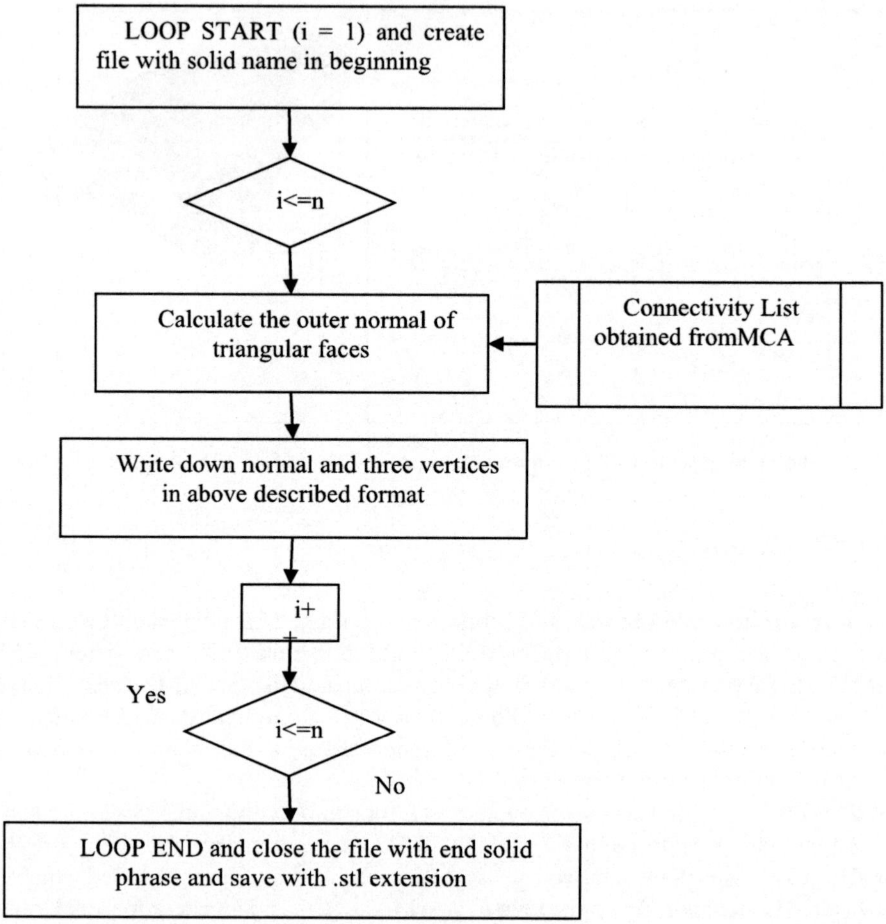

Fig. 5 Flow chart of STL file generation from Matrix D

results suggested that the dependency on software can be minimized to produce STL from point cloud. In future, the study may be extended to reduction of computational time by applying data segmentation on point cloud. Moreover, the data segmentation and filling region can be done by integrating MCA with advance algorithm of STL repairing and point reducing.

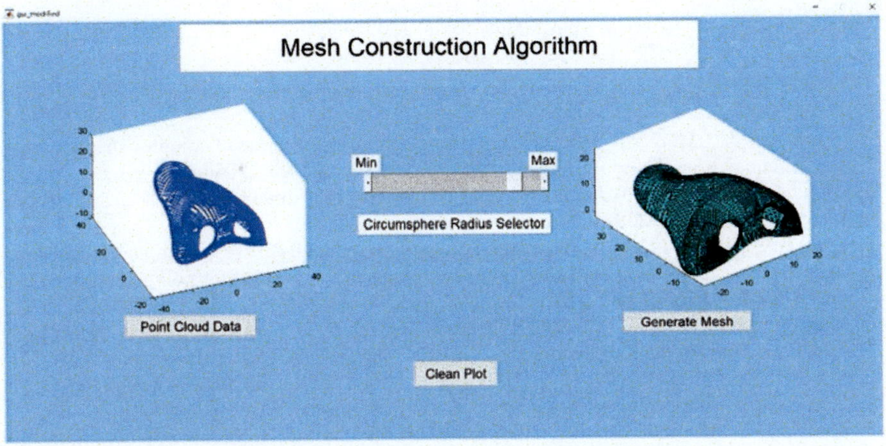

Fig. 6 Developed graphical user interface (GUI)

Table 1 Printed samples and processing time

Input Points	Part Volume (mm^3)	Point Cloud	3D printed part	Time (seconds)
206,682 (Bone)	4 × 13 × 4			619
270,036 (Human Nose)	45 × 60 × 27	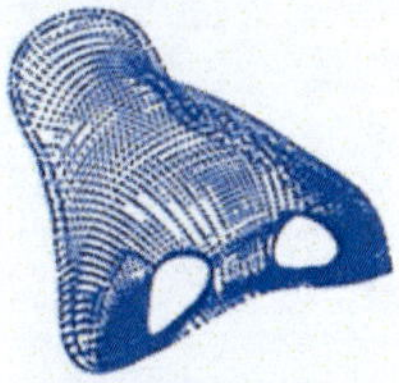	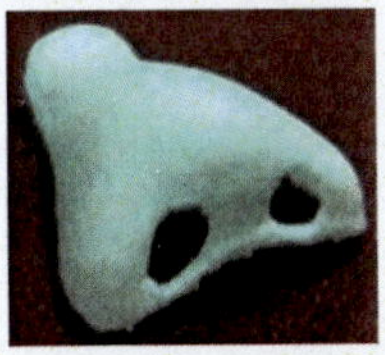	812
100,926 (Pot)	45 × 45 × 40	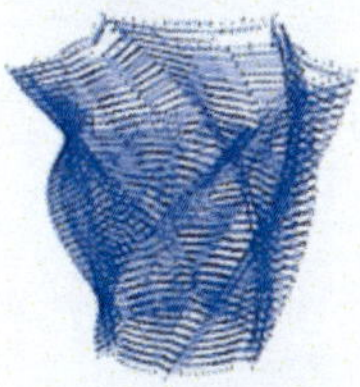		156
151,392 (Sunglass)	32 × 17 × 10			337

References

1. Choi BK, Shin HY, Yoon YI et al (1988) Triangulation of scattered date in 3D space. Comput Des 20:239–248
2. Boissonnat J-D, Geiger B (1992) Three dimensional reconstruction of complex shapes base on the Delaunay triangulation. Biomed Image Process Biomed Vis 1905:964–975
3. Várady T, Martin RR, Cox J (1997) Reverse engineering of geometric models—an introduction. Comput Des 29:255–268
4. Piegl LA, Tiller W (2000) Surface approximation to scanned data. Vis Comput 16:386–395
5. Woo H, Kang E, Wang S et al (2002) A new segmentation method for point cloud data. Int J Mach Tools Manuf 42:167–178
6. Chen YH, Ng CT, Wang YZ (1999) Generation of an STL File from 3D measurement data with user-controlled data reduction. Int J Adv Manuf Technol 15:127–131
7. Lee SH, Kim HC, Hur SM et al (2002) STL file generation from measured point data by segmentation and Delaunay triangulation. CAD Comput Aided Des 34:691–704
8. Christiansen HN, Sederberg TW (1978) Conversion of complex contour line definitions into polygonal element mosaics. Comput Graph (ACM) 12:187–192
9. Xu J, Hou W, Zhang H (2018) An improved virtual edge approach to slicing of point cloud for additive manufacturing. Comput Aided Des Appl 15:330–336
10. Oropallo W, Piegl LA, Rosen P et al (2018) Point cloud slicing for 3-D printing. Comput Aided Des Appl 15:90–97
11. Yuan T, Peng X, Zhang D (2018) Direct rapid prototyping from point cloud data without surface reconstruction. Comput Aided Des Appl 15:390–398
12. Jinting Xu, Hou W, Sun Y, Lee Y-S (2018) PLSP based layered contour generation from point cloud for additive manufacturing. Robot Comput-Integrated Manuf 49:1–12

Sustainable Green Lean Six Sigma Methodology and Application Status: A Perspective Review

Vishwas Yadav, Pardeep Gahlot, Mahender Singh Kaswan, Rajeev Rathi, and Mahipal Singh

Abstract Green lean six sigma (GLS) is an all-inclusive methodology, assists in optimal usage of resources, minimizes wastes, and dispenses a pathway for sustainable development. The progressive manufacturing sectors are in search of approaches that will validate them to be sustainable along with yield competitive assistance. Green, lean, and six sigma strategies are some of the extensively employed tools which assist the organizations towards sustainability. The objective of this manuscript is to review the literature on GLS strategy, methodology, and successful implementation in various sectors like manufacturing, food processing, and construction, automobile, mine, and public sectors. A qualitative kind of review methodology was embraced to carry out the review. Research articles associated with GLS strategy were acquired from well-reputed journal databases. Almost thirty-five research articles from the previous seven years related to the GLS approach have been taken for this purpose. The analysis shows that a framework for the execution of GLS was instigated. The proposed GLS framework will help the manufacturing sectors through the mitigation of rejections, CO_2 emissions, and imperfections in the manufacturing processes. This review manuscript emphasizes the perception of green lean six sigma by carrying out a methodical review concerning all sectors like manufacturing, food processing, construction, automobile, mine, and public sectors and the development of a framework which is equally important from a sustainability point of view. This manuscript assists the researchers to execute GLS strategy to achieve functional and operational benefits.

Keywords Green lean six sigma · Sustainability · Framework · Lean six sigma · Six sigma

---

V. Yadav · P. Gahlot
Department of Mechanical Engineering, MaharashiDayanad University, Rohtak, India

M. S. Kaswan (✉) · R. Rathi · M. Singh
School of Mechanical Engineering, Lovely Professional University, Phagwara, India

A. K. Dubey et al. (eds.), *Recent Trends in Industrial and Production Engineering*, Lecture Notes in Mechanical Engineering,
https://doi.org/10.1007/978-981-16-3135-1_26

1 Introduction

In recent days, each organization aims at profitability, efficiency, quality, responsiveness, customer satisfaction, and sustainability of services to protract in this era of competition in the market. The customers demand the product of high quality with minimum environmental waste, at the reasonable cost, and available with the minimum time frame [16, 26]. To meet the customer demand and governmental policies on the climate mitigation industry has to use a minimum quantity of resources, with lesser emission and societal damage. The industrial organization through the adoption of sustainable methods can take a competitive edge over the market through the production of eco-friendly, high-quality products. This will not only increase organizational competitiveness but also bring social equity and a healthy economic perspective. Continuously enhancing environmental studies have transferred the centre of attention of manufacturing sectors from conventional points, quality, a satisfaction of customers to sustainability [1, 27, 11]. Sustainability is not only to esteem the bottom line of 3P (people, plant, and profit) but also, it is the accordance of assets to face people's needs [28]. Moreover, the manufacturing sectors pay more attention to social, economic, and institutional sustainability, and it will expend more and target green actions [37]. The customer desires to buy the products that are of topmost quality, easily accessible at lower costs, and more preferable is available on time. Lean green strategy stresses on the minimal exploitation of resources through waste minimization and mitigates the harmful environmental effects through the minimization of hazardous gases. Thus, the integrated green lean strategy assists in economizing the funds, better environmental situations, and prefers social equity in the manufacturing firms as a whole. Figure 1 depicts a demand-driven model for sustainable results. There are many quality improvement approaches like statistical quality control (SQC), total quality management (TQM), and green lean six sigma (GLS) which were embraced by the organizations for sustainable improvement [46]. Out of these approaches, GLS is adduced as a budding, most effectual continuous enhancement approach. As, GLS is a comprehensive perspective that

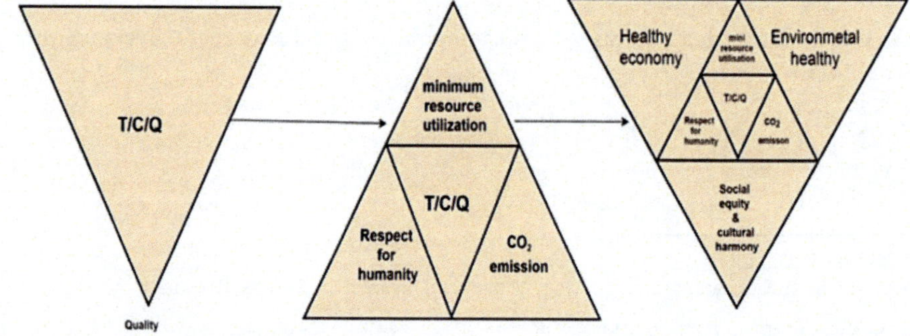

Fig. 1 Demand-driven sustainable model

minimizes the rejection along with a reduction in variation in the particular process and negative environmental impact with the help of the green concept of 3'R (recycle, reuse, and reduce) [6]. The main objective of this approach is better exploitation of natural resources, minimizes waste generation, and imparts a way for sustainable development.

GLS encompasses three different approaches that are green, lean, and six sigma to enhance productivity, profitability, manufacturing process, cycle time, and sustainability [22]. The key objective of lean manufacturing is to exhibit high-quality products or services at the bottommost cost and in the minimum time by mitigating rejections [7, 25]. The key objective of the green concept is to minimize the environmental impact of manufacturing, diminishing energy, and pollution effect of the product through consumption and recycling materials; subsequently, the product is rejected [36]. In the 1980s, six sigma was initiated by Motorola, and it was completely executed in 1988 [14]. The main goal of this concept is to modify the performance of a process and attain high-quality products or services by evaluating and abolishing the root causes of defects and reducing variation in the processes [49]. But, integrated GLS recommends a process or product that is of the best quality, cost, and eco-friendly on an account of minimization in rejection, defects, and negative environmental impacts of the product [22].

In recent days, admiration of GLS concept is increasing all over the world. Each manufacturing sector's interest is also enhanced to get full knowledge about GLS and its methodology to execute in their ventures. From the literature survey, it was cleared that most of the work was associated with the direct execution of GLS strategy. But still, the literature necessitates provision on how to pertain GLS? What should be the best methodologies and tools for all the phases? Therefore, the main objective of the present study is to perquisite GLS concepts, methodology, and execution. The current study also highlights future research guidance in this field.

2 Integration Model of Green and Lean Six Sigma

The sustainable manufacturing sector must carry out a robust on the environmental aspects. There is a great need for all manufacturing sectors to re-evaluate their entrepreneurial activities due to increased perception about sustainability and requisition for environmentally sound products [39]. Therefore, manufacturing sectors are investing more funds for the adoption of sustainable practices of manufacturing and consumption [2, 8]. The history of GLS pursued the enlargement of the lean initiative. Lean recommends for the structured exclusion of rejections over excellence at entirely stages contained by the manufacturing firms [30, 48]. The lean method minimizes rejections but it is not capable to perceive the variation in the process together mitigation in environmental impacts. Thus, the requirement of green and six sigma concept was detected to produce environmentally sound products of the higher stipulation. Green concept mitigates the harmful effects of the by-products by forming it more environmentally sound [6, 13]. But this concept is not able to detect process variation. Thus, there is a requirement of six sigma strategy to fabricate the processes with less variation. Six sigma approach minimizes process variation that assists to reduce waste of the products [24]. GLS strategy has been adopted in organizations to create a sustainable environment and to manufacture high specification products at lower costs [12]. But a single strategy is not able to convey all the factors completely associated with sustainability [29]. Therefore, there is a great need to integrate GLS strategy that minimizes the rejection along with the reduction in variation and harmful ecological effects [17]. Also, integrated green lean six sigma strategy effectively produces high specification and eco-friendly products at the bottommost cost [15]. Therefore, incorporation of green, lean, and six sigma can be considered as the latest imminent to the manufacturing sectors for enhancement in sustainability (Fig. 2).

3 Proposed GLS Implementation Framework

Define-measure-analyse-improve-control (DMAIC) cycle is the methodology which is employed to carry out six sigma approach successfully for any venture for getting better business processes to reduce defects [10]. DMAIC cycle could dispense green lean with a explicit and comprehensive project-employed initiation to the execution and accomplishment of green strategies, which can be modified over an account of the Kaizen approach. Therefore, green approaches are executed, directed, assisted, and modified under the umbrella of six sigma's DMAIC methodology. So, this cycle helps extract and affix six sigma and lean's practices and tools throughout DMAIC's five phases when accompanying green projects. Lean and six sigma will also provide some specific techniques and tools which will be helpful to recognize define, prioritize, control, govern, protract, and modify green strategies. DMAIC cycle is the crucial methodology for directing GLS approach. Each phase of this cycle was explained as follows:

Define: This phase is very pre-eminent because it is imperative to define a suitable job for process enhancement. This phase is also used to recognize and adopt the right project additionally illuminate the scope of the project [11, 34]. The requirements, priorities, and prospects of the business and customers are intimated as voice of business (VOB) and voice of customers (VOC). The define phase helps manufacture sectors to prioritize green strategies and assets as a means to decide which organizational activity should be employed firstly and in respect of whichever variable (i.e. water consumption, energy consumption, CO_2 emissions, etc.) should be focused. Manufacturing, facilities, logistics, marketing, process, and product design, etc., are included in organizational activity/function. Six sigma recommended various specified tools which include project ranking matrix, Pareto analysis, project selection matrix, quality function deployment, Pareto priority index, analytical hierarchy process (AHP), and theory of constraints, etc.

Measure: The purpose of the measure phase is to explain the latest ceremonial of the structure and to establish the crucial aspects and matrices concerning quality

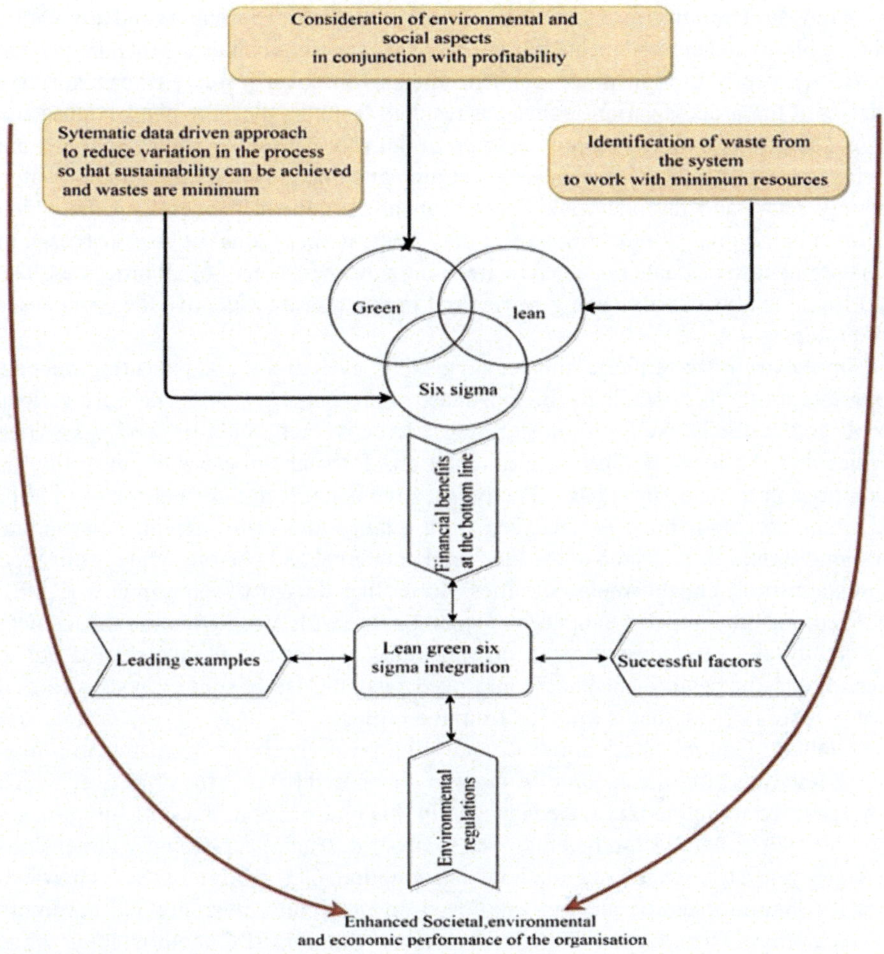

Fig. 2 Green lean six sigma model

and environment [21, 33]. Further, the target of this phase is to authenticate the root causes of the problem. In this phase, an environmental current state map has been established to dispense an exhaustive perception of the latest manufacturing process. Environmental value stream mapping (EVSM) embraces details concerning power consumption, material consumption, water consumption, and environmental effects besides traditional value stream mapping specifications like cycle time, change over time, uptime, and lead time for every practices. Environmental current state mapping development assists to visualize the genuine manufacturing trends and dispenses awareness on resource consumption from the environment. The measure phase is used to fulfil the objectives as established in the define phase.

Analyse: The purpose of analyse phase is to identify the root causes and to examine the problems and method incapability. This phase specifies enhancement pursuits that must be executed to attain process enhancements. Moreover, it studied the prospective causes of the process variation that was resulted from the measure phase. Cause and effect diagram is widely employed to point out root causes. Five-way analyses are being employed to detect the waste generation and uncontrolled utilization of water, energy, CO_2, and raw material. These tools are beneficial to identify gaps which furnish in the present operations from the foremost operations. LCA is also used in this segment to evaluate prospective green influences on behalf of all processes. The statistical analysis is also being carried out in this phase to identify the root causes of problems.

Improve: Improve phase aims at suggesting, evaluating, and executing the best possible solutions to eradicate the root cause of problems [11, 32]. There are various tools and methods like 5S, brainstorming, corrective action matrix, and poka-yoke which is used to determine the best solutions for managing waste generation as point out in the measure phase. The perspective benefits and results achieved after efficient execution must be analysed from quality and environment. Appropriate training should be provided to the industrial personnel and persons who are engaged in establishing enhancement activities. Tools like design of experiments (DOE), Kaizen, and other process enhancement methods are widely used for the enhancement in quality and productivity of an organization. A pilot test is also carried out to accomplish the performance of the improved design. Cost–benefit analysis is carried out to assess saving that is initiated from the project.

Control: Control phase aims at monitoring properly the performance improvement activities after the accomplishment of concerned [9]. All the objectives which are specified in define phase are achieved in this phase. Then, the statically process control (SPC) method can be employed to observe crucial eco-friendly dimensions such as water, energy and raw material consumption, CO_2 emissions. SPC chart will assist to monitor the parameters employed on maximum, average, and minimum consumption. Then, remedial steps were taken through SPC methodology, when the energy, water, and raw material consumption or CO_2 emissions surpasses the extreme predefined boundary. SPC will assist in the exploration and understanding of the level and causes of variation in consumption [20]. This will also help organizations in decision-making and accommodating the detection of difficulties. All enhancement activities are monitored properly but the control phase of DMAIC cycle will be suggested for the best improvement activities. It also motivates all the employees of the organization for the minimization of variability in terms of consumption of green parameters. The proposed framework applies to almost all sectors like manufacturing, food processing, mine, public sectors, construction, and automotive sectors. But, the analysis reveals that manufacturing sectors will be more benefited through the execution of this framework (Fig. 3).

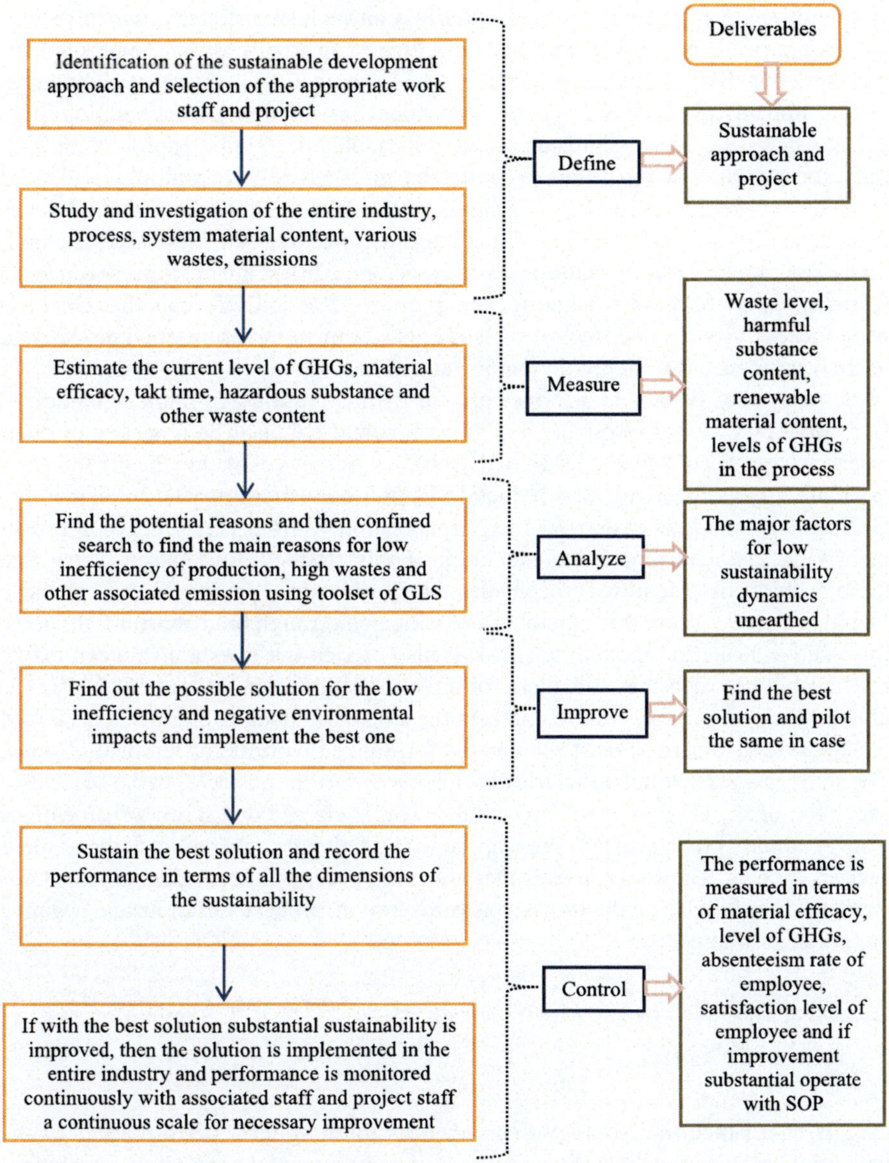

Fig. 3 GLS implementation framework

4 Current Status of GLS Implementation

The current manuscript is targeted on GLS strategy, methodology, and successful implementation in various sectors like manufacturing, food processing, construction, automotive, mine, and public sectors. GLS is an inclusive strategy, which recommends a process or product that is of the highest quality, cost, and environment-friendly over the minimization in rejections, imperfections, and negative environmental impacts [22]. GLS is greatly dependent on six sigma's methodology that is DMAIC cycle. This cycle is employed to establish six sigma approach successfully for any project for obtaining better business processes to minimize defects [10]. Table 1 depicts green lean six sigma methodology and implementation status in various sectors. The purpose of this paper is to critically review GLS methodology and implementation status in various sectors. Almost the previous seven years research papers have been taken for this purpose. The study reveals that GLS has been successfully applied to almost all sectors like manufacturing, food processing, health care, construction, public sectors, and mines.

Some articles presented a framework for the integration and implementation of GLS strategy to improve quality and environmental effects. The objective of other research papers is to develop DMAIC-based GLS framework within the organization to obtain green objectives for a reduction in environmental impacts. Implementing GLS strategy is not an easy task in any organization. Various performance improvement GLS enablers hinder the execution of this strategy. Rest papers show that these enablers are identified to consult with the industrial personnel and extensive literature survey. Then, these enablers are validated through brainstorming sessions. Interpretive structural modelling (ISM) is also carried out to establish a contextual relationship between these enablers for the removal of them. MICMAC analysis is also employed to classify these enablers for a better understanding of them. BWM technique is also carried out to screen out top-ranked barriers for removal of them. The study reveals that manufacturing sectors faces various obstacles in the successful execution of the GLS strategy. From Table 1, it is cleared that if any organizations want to employ green lean six sigma approach effectively, they require to target these success aspects. The study reveals that after the successful implementation of GLS strategy, there will be a reduction in waste generation, process variation, and negative environmental impacts.

5 Research Gaps

GLS is an all-inclusive strategy that enhances the long-term performances of the industrial organizations over the execution of environmental assessments, manufacturing of eco-friendly products, and dispenses customer-targeted top quality products. The manufacturing sectors are forced to embrace sustainable practices like GLS due to excessive pressure of governmental policies. Therefore, humanity can be exploited through minimum environmental deterioration in terms of minimized rejections and CO_2 emissions through proper execution of GLS strategy. But, in the literature, there is no evidence of assortment of cases in which the organizational culture is not accessible to green measures. The study must be carried out for understanding the function of organizational culture in green strategies. There is no evidence to

Table 1 Green lean six sigma methodology and implementation status in various sectors

References	Sector	Remarks
[47]	Automotive sector	The aim of the article is to establish the correlation among GLS and management innovation (MI). This paper also attempts to explore how GLS can be promoted and has a pragmatic effect on the establishment of MI to attain superior performances in automotive sector. This results in a reduction in waste generation, process variation, and environmental impacts in this sector
[10]	Manufacturing sector	The purpose of the article is to provide a theoretical model for the integration of green, lean, and six sigma to enhance its effectiveness. This paper also suggests DMAIC methodology which is used to overcome the drawbacks of green lean
1	Construction sector	This article presents a framework and assimilates green, lean, and six sigma in a logical strategy with the objective of enhancing the quality and environmental effects of the construction process. This GLS framework furnishes an inclusive, multi-phase strategy for the enhancement in process, and reduction in the life cycle environmental impacts in the construction process
[13]	Manufacturing sector	The main objective of the manuscript is to pre-eminent review the green lean strategies, highlight the drawbacks, and assess the compatibility of the lean, green, and six sigma concepts. This paper also presents DMAIC methodology of six sigma approach to show that systematic literature review (SLR) specifies that integration of green lean may have assumed similar drawbacks as the independent green and lean strategies, but these may be improved on with the help of GLS approach
[4]	Manufacturing sector	This paper confers a framework that systematically directs manufacturing sectors over a five-phase and sixteen steps process to assimilate and execute GLS strategies successfully to enhance the performance of sustainability. The study reveals that averagely resource consumption and cost of energy or mass stream reduce from 20 to 40% and 7–12%, respectively, after the execution of GLS strategy
[22]	Automobile sector	The purpose of this paper is to identify barriers in GLS product development (GLSPD) from the consultation with the industrial personnel and extensive literature review. Twenty-one barriers are identified and validated by the brainstorming sessions. Then, interpretive structural modelling (ISM) has been used to instigate a hierarchical model of these barriers in executing GLSPD process in the automobile sector. Further, MICMAC analysis was also carried out to categorize for the better understanding of these barriers

(continued)

Table 1 (continued)

References	Sector	Remarks
[35]	Manufacturing sector	This paper presents the successful implementation of lean six sigma framework with environmental considerations to assure the reduction in defects and environmental impacts. The study reveals that internal defects have been reduced from 16,000 to 5000 PPM. Environmental impacts were also minimized from 42 to 33 Pt
[5]	Manufacturing sector	The purpose of this article is to identify GLS barriers and to establish the contextual relationship between these barriers. Systematic literature review (SLR), interpretive structural modelling (ISM), and MICMAC analysis were also carried out for the removal of these identified barriers. Fifteen barriers were identified that hinder the execution of GLS strategy. This study can also help the industrial managers for a better understanding of these barriers to a suitable economic and environmental exhibition
[31]	Food industries	This paper highlights the application of lean six sigma and assesses the effect of LSS on environmental sustainability in food processing industries. This results in the reduction in process variation, rejections, and environmental impacts. This paper also proposed VSM-DMAIC methodology
[48]	Health care sector	This paper presents a systematic framework based on a traditional DEF meta modal, which prescribes particular aspects and synergies among lean and green approaches. This framework also assists more inspection of green supply chain in hospitals. These results in a reduction in the waste generation in the hospital operations and successfully related greening to the lean abilities may permit to reduce environmental impacts within the hospitals
[3]	Service industries	This article essentially reviews LSS methodology and emphasizes their significance to attain sustainable services. This paper also proposed a framework to identify nine critical factors, which are hindering the execution of LSS approach through sustainability. These results provide a track to obtain equity in economic, social, and environmental grades in service industries
[29]	Manufacturing sector	This paper proposes GLS strategies to identify and rank performance improvement enablers. Eighteen enablers and five enablers' dimensions are screened out from the expert's opinion and extensive literature survey. Then, AHP technique is carried out to rank these enablers and dimension enablers

(continued)

Table 1 (continued)

References	Sector	Remarks
[45]	Public sector	The purpose of this paper is to provide a green lean six sigma modal for public sectors. This modal enhances the functional brilliance of public sector assistance like telecommunication, construction, and health care. GLS is integrated with supply chain management to remove obstacles withstand by public sectors, leading to process enhancement. The study reveals that the proposed modal can also apply to other public service organizations
[17]	Manufacturing sectors	The objective of this article is to identify GLS enablers, and modelling is done with the help of interpretive structural modelling (ISM) and methodically determines the relationship between these enablers. Further, impact matrix cross-reference multiplication applied to a classification (MICMAC) analysis has been employed to categorize these enablers for superior interpretation. The study reveals that twelve GLS enablers, which are hindering the execution of GLS strategy, are identified and modelled
[44]	Mine sector	The aim of the paper is to propose a framework for executing GLS approach within the organization to achieve the green objective through DMAIC process of the five aspects of the environmental exhibition. In this manuscript, a case study approach is employed to execute GLS framework to minimize the level of graphite and dust pollution in cast mine
[39]	Manufacturing SMEs	This paper presents a conceptual framework for combined green lean strategy and sustainability for small-medium enterprises. The outcome specifies that most general provocation to green lean execution is a shortage of matrices and evaluation. The outcomes are beneficial for SMEs possessor, academic persons, and industrial managers to accept measures for the improvement in sustainability
[15]	Construction sector	The objective of this manuscript is to analyses barriers to the GLS construction process through the literature survey and opinion of the experts. Further, these barriers were validated through brainstorming sessions. IMS methodology was also carried out to create 11 level hierarchical modal. MICMAC analysis is also used to classify these barriers for better understanding. The study reveals that there will be a reduction in wastage generation process variation and environmental impacts after the execution of GLS strategy in the construction industry

(continued)

Table 1 (continued)

References	Sector	Remarks
[19]	Manufacturing sector	The aim of this article is to integrate and develop DMAIC-based GLS framework. The presented framework dispenses a pathway for GLS execution through a suitable preference for the project. The study reveals that many sustainability measures have been estimated with the help of unique GLS indices and toolset
[38]	Manufacturing sector	The purpose of this paper is to execute green approach into lean six sigma approach projects in manufacturing sectors. This article also recognizes under which circumstances the output of LSS projects with prudent benefits are more environmentally sound. This results in the achievement of greener, customized, and economic-oriented outcomes
[19]	Manufacturing sector	The purpose of this article is to prioritize performance improvement GLS enablers with the help of best-worst method (BWM). Twelve enablers are identified to consult with the industrial personnel. Then, the top three ranked enablers are screened out of these twelve enablers through BWM and further justified through some advanced decision-making approaches like AHP, DEMATEL, and PROMETHEE-II
[43]	Manufacturing sector	This article comprises the ranking of environmental LSS enablers with the help of best-worst method (BWM). Out of thirty enablers, twenty-two were finalized on the basis of expert's opinion and statistical analysis. The results reveal that the top-ranked enablers were beneficial for the effective execution of 'qweco-friendly LSS in any manufacturing sector. Further, the results are validated through AHP and ANP

create a questionnaire for exploration of the interrelation among GLSS operations and managerial innovation in automobile sectors. Additionally, there is no evidence in the literature that developed framework can be carried out for more manufacturing and industrial organizations where necessitate employing LSS initiatives combined with ecological issues was insistent. It can be attained by additional improvement in the current stages and proposing different tools and methods in all steps to get consistent outcomes through the developed framework. Also, the literature lacks to prioritize and ranking GLS enablers by fuzzy BWM technique and further validated by other advanced decision-making strategies like fuzzy TOPSIS and fuzzy PROMETHEE. Moreover, no evidence in the literature shows the significance of LCA in product enlargement and its prospective impact on the selection of a particular project for sustainability. Further, there is no evidence for the development of hybrid strategy of imperishable LSS execution.

6 Conclusion and Future Research Agenda

The present work carried out that is associated with various sectors like manufacturing, food processing, mine, public sectors, construction, and automotive. Green lean six sigma demonstrated to be a real business approaches that provide allover enhancement at the level of the process. GLS has been renowned as an all-encompassing strategy that diminishes harmful environmental effects and furnishes a better quality of product simultaneously. The efficiency of the organizations has been improved after the successful implementation of GLS strategy. GLS framework will facilitate the organizations for the reduction in waste generation, process variation, and harmful environmental impacts. Averagely resource consumption and cost of energy or mass stream also reduced after the successful implementation of GLS strategy. The current study can also help the industrial managers for a better understanding of GLS barriers for suitable economic and environmental performance. The study also reveals that internal defects and environmental impacts were also reduced significantly. GLS methodology dispenses a path to obtain equity in economic, social, and environmental grades in service industries. GLS framework also reduces life cycle environmental impacts in the construction process. The organizations must employ it through the accurate methodology and take advantage of appliances and techniques for the effectual use of GLS approach. The sectors can get remarkable benefits by appropriate implementation of this strategy. To execute GLS methodology to their organizational units with greater perception, the current study will disseminate manufacturing sectors. This present study also motivates the various sectors to target on target on success factors that are significant for the success of this approach. The most of published research on the application of green and lean operations has confessed about them as an interdependent and executed strategy. In the future scope in this field, scholars and practitioners may use the application of LCA in product development and its inherent impacts on the selection of a particular project for sustainable development. Further, this study also motivates them to target the role of GLS for sustainability improvement through industry 4.0 and modelling and exploration of its performance improvement barriers. Moreover, this study animates the scholars to expect, cause, and employ sustainability to protect the environment, obtain long-term outcomes, and pertinently care for their personnel and community. Further, scholars and practitioners may think about the study of the applications of GLS methodology to the service initiatives. Moreover, if the employees of the organizations are well skilled with the procedure, this can be smoothly executed, and outcomes will reveal at the basis of supply chain management (SCM).

References

1. Banawi A, Bilec MM (2014) a framework to improve construction processes: integrating Lean, Green, and Six Sigma. Int J Constr Manag 14(1):45–55
2. Bond A, Dusik J (2019) Impact assessment for the twenty-first century-rising to the challenge. Impact Assess Proj Apprais 38(2):126
3. Caiado R, Nascimento D, Quelhas O, Tortorella G, Rangel L (2018) Towards sustainability through green, lean and six sigma integration at service industry: review and framework. Technol Econ Dev Econ 24(4):1659–1678 ISSN: 2029-4913
4. Cherrafi A, Elfezazi S, Chaiarini A, Mokhlis A, Benhida K (2016) The integration of lean manufacturing Six Sigma and sustainability: a literature review and future research directions for developing a specific model. J Clean Prod 139:828–846
5. Cherrafi A, Elfezazi S, Govindan K, Garza-Reyes JA, Benhida K, Mokhlis A (2017) A framework for the integration of green and lean six sigma for superior sustainability performance. Int J Prod 55(15):4481–4515
6. Deif AM (2011) A system model for green manufacturing. J Clean Prod 19(14): 1553–1559
7. Denis P (2007) Lean production simplified. Productivity Press, New York
8. Duarte S, Cruz-Machado V (2019) Green and lean supply-chain transformation: a roadmap. Prod Plan Control 30(14):1170–1183
9. Forrester PL, Shimizu UK, Soriano-Meier H, Garza-Reyes JA, Cruz Basso LF (2010) Lean production, market share and value creation in the agricultural machinery sector in Brazil. J Manuf Technol Manage 21(7):853–871
10. Garza-Reyes JA, Fliant JA, Kumar V, Antony J, Soriano-Meier H (2014) A DMAIC approach to lead time reduction in an aerospace engine assembly process. J ManufTechnolMana 25(1):27–48
11. Garza-Reyes JA, Oraigige I, Soriano-Meier H, Harmanto D, Rocha-Lona L (2010) An empirical application of six sigma and DMAIC methodology for business process improvement. In: Proceedings of the 20th international conference on flexible automation and intelligent manufacturing (FAIM), San Francisco, CA, US, July 12–14
12. Garza-Reyes JA, Yu M, Kumar V, Upadhyay A (2018) Total quality environmental management: adoption status in the Chinese manufacturing sector. TQM J 30(1):2–19
13. Garza-Reyes JA (2015) Green lean and need for Six Sigma. Int J Lean Sigma 6(3):226–248
14. Goh TN (2002) A strategic assessment of six sigma. Qual Reliab Eng Int 18(5):403–410
15. Hussain K, He Z, Ahmad N, Iqbal M (2019) Green, Lean, Six Sigma barriers at a glance: a case from the construction sector of Pakistan. Build Environ 161:106225
16. Kapur S, Kaswan MS (2020) Ergonomic assessment of the lifting tasks performed by North Indian workers in LPG cylinder distribution supply chain. In: International conference on applied human factors and ergonomics. Springer, Cham, pp 252–258
17. Kaswan MS, Rathi R (2019) Analysis and modeling the enablers of green lean six sigma implementation using interpretive structural modeling. J Clean Prod 231:1182–1191
18. Kaswan MS, Rathi R (2020) Green Lean Six Sigma for sustainable development: integration and framework. Environ Impact Assess Rev 83:106396
19. Kaswan MS, Rathi R (2020) Investigating the enablers associated with implementation of Green Lean Six Sigma in manufacturing sector using Best Worst Method. Clean Technol Environ Policy, 1–12
20. Kaswan MS, Rathi R, Khanduja D, Singh M (2020) Life cycle assessment framework for sustainable development in manufacturing environment. In: Advances in intelligent manufacturing. Springer, Singapore, pp 103–113
21. Kaswan M, Rathi R, Singh M (2019) Just in time elements extraction and prioritization for health care unit using decision making approach. Int J Qual Reliab Manage 36(7):1243–1263
22. Kumar S, Luthra S, Govindan K, Kumar N, Haleem A (2016) Barriers in Green Lean Six Sigma product development process: an ISM approach. Prod Plan Control 27(7–8):604–620
23. Kumar M, Kaswan MS (2016) Optimization of surface roughness & MRR in end milling On D2 steel using Taguchi method. Optimization 5(1)

24. Kumaravadivel A, Natarajan U (2013) Application of Six-Sigma DMAIC methodology to sand-casting process with response surface methodology. Int J Adv Manuf Technol 69(5–8):1403–1420
25. Liker JK (1996) Becoming lean. Free Press, New York
26. Marousek J, Haskoya S, Zeman R, Vanickova R (2015) Managerial preferences in relation to financial indicators regarding the mitigation of global change. Sci Eng Ethics 21(1):203–2017
27. Mollenkopf D, Stolze H, Tate WL, Ueltschy M (2010) Green, lean, and global supply chains. Int J Phys Distrib Logist Manag 40(1/2):14–41
28. Nunes B, Bennett D (2010) Green operations initiatives in the automotive industry: an environmental reports analysis and benchmarking study. Benchmark: Int J 17(3):396–420
29. Pandey H, Garg D, Luthra S (2018) Identification and ranking of enablers of Green Lean Six Sigma implementation using AHP. Int J Product Qual Manag 23(2):187–217
30. Panwar A, Jain R, Rathore APS (2018) Obstacles in lean implementation in developing countries-some cases from the process sector of India. Int J Lean Enterprise Res 2(1):26–45
31. Powell D, Lundeby S, Chabada L, Dreyer H (2017) Lean six sigma and environmental sustainability: the case of a Norweign dairy producer. Int J Lean Six Sigma 8(1) ISSN: 2040-4166
32. Rathi R, Khanduja D, Sharma SK (2015) Synergy of fuzzy AHP and Six Sigma for capacity waste management in Indian automotive industry. Decis Sci Lett 4(3):441–452
33. Rathi R, Khanduja D, Sharma SK (2016) A fuzzy MADM approach for project selection: a Six Sigma case study. Dec Sci Lett 5(2):255–268
34. Rathi R, Khanduja D, Sharma SK (2017) A fuzzy-MADM based approach for prioritizing Six Sigma projects in the Indian auto sector. Int J Manage Sci Eng Manage 12(2):133–140. https://doi.org/10.1080/17509653.2016.1154486
35. Ruben R, Vinodh S, Asokan P (2017) Implementation of Lean Six Sigma framework with environmental considerations in an India automotive component manufacturing firm: a case study. Prod Plan Control. https://doi.org/10.1080/09537287.1357215
36. Russell-Smith SV, Lepech MD, Fruchter R, Meyer YB (2015) Sustainable target value design: integrating life cycle assessment and target value design to improve building energy and environmental performance. J Clean Prod 88:43–51
37. Saunila M, Ukko J, Rantala T (2018) Sustainability as a driver of green innovation investment and exploitation. J Clean Prod 179:631–641
38. Shokri A, Li G (2020) Green implementation of Lean Six Sigma projects in the manufacturing sector. https://www.emerald.com/insight/2040-4166.htm
39. Siegel R, Antony J, Garza-Reyes JA, Cherrafi A, Lameijer B (2019) Integrated green and lean approach and sustainability for SMEs from a literature review to a conceptual framework. J Clean Prod 240:118205
40. Singh M, Rathi R (2019) A structured review of Lean Six Sigma in various industrial sectors. Int J Lean Six Sigma. https://doi.org/10.1108/IJLSS-03-2018-0018
41. Singh M, Kumar P, Rathi R (2019) Modelling the barriers of Lean Six Sigma for Indian micro-small medium enterprises: an ISM and MICMAC approach. TQM J. https://doi.org/10.1108/TQM-12-2018-0205
42. Singh M, Rathi R, Garza-Reyes JA (2020) Analysis and prioritization of Lean Six Sigma enablers with environmental facets using best worst method: a case of Indian MSMEs. J Clean Prod 279:123592
43. Singh M, Rathi R, Khanduja D, Phull GS, Kaswan MS (2020) Six Sigma methodology and implementation in Indian context: a review-based study. In: Advances in intelligent manufacturing. Springer, Singapore, pp 1–16
44. Sony M, Naik S (2019) Green Lean Six Sigma implementation framework: a case of reducing graphite and dust pollution. Int J Sustain Eng 13(3):1–10
45. Sreedharan VR, Raju R (2018) Development of a Green Lean Six Sigma model for a public sector. Int J Lean Six Sigma 9:238–255
46. Wang H (2008) A review of Six Sigma approach: methodology, implementation and future research. In: 4th international conference on wireless communication, networking and mobile computing, 2008. WiCOM'08. IEEE, New York, pp 1–4

47. Zamari SN, Habidin NM, Hibabadullah NF, Fuzi NM, Desa AFNC (2013) Green Lean Six Sigma and managerial innovation in Malaysian automotive industry. Bus Mang Strategy 4(1):97
48. Zhu Q, Johnson S, Sarkis J (2018) Lean six sigma and environmental sustainability: a hospital perspective. Supply Chain Forum Int J 19(1):25–41
49. Zu X, Fredendall LD, Douglas TJ (2008) The evolving theory of quality management: the role of Six Sigma. J Oper Manag 26(5):630–650

The Effect of Gating System on Quality of Traditional Rural Metal Castings of India

Soumyajit Roy, Akshay Kr Pramanick, and Prasanta Kr Datta

Abstract Wax-based investment casting is a popular technique to produce complex-shaped products that require very little machining for use. Experiments were conducted for understanding the effect of the gating system on dimensional accuracy as well as the metal quality of cast products having different thickness in case of restricted and unrestricted sections. The production technique is followed by traditional rural artisans of India. The casting method uses bee's wax as pattern material and local clay as the mold material. 60/40 brass is generally melted, and the liquid metal is poured in hot clay molds. For gradation of cast samples, technique for order of preference by similarity to ideal solution (TOPSIS) was used for evaluation.

Keywords Wax-based investment casting · Gating system · Hot clay mold · Indian traditional rural casting method · 60/40 brass · TOPSIS

1 Introduction

Investment casting or near-net-shape casting process is used to produce complex-shaped or thin-walled cast products with close-dimensional tolerances [1]. Little machining is required for these types of components. This process is known as "Dhokra" casting, a rural casting technique, famous for hollow casting production, and can be found in West Bengal, Orissa, Chhattisgarh, and Jharkhand of India. In this process, pattern materials contain a synthetic alloy of bees wax and sal-dammer resin, and mold materials use locally available lateritic clay with fine silica sand.

Kinetics of liquid metal [2] inside a mold controls the quality of casting, and so, this becomes of primary importance for thin casting sections (like "hollow Dhokra" castings). The gating system and the position of the gate also control the flow. This demands additional care to select the proper gating parameters for better quality as the liquid metal flow gets resisted by high frictional drag (of large flat mold surface) as well as the capillary action due to too narrow mold cavity. Although, when liquid

S. Roy (✉) · A. Kr Pramanick · P. Kr Datta
Metallurgical and Material Engineering Department, Jadavpur University, Kolkata, India

A. K. Dubey et al. (eds.), *Recent Trends in Industrial and Production Engineering*, Lecture Notes in Mechanical Engineering,
https://doi.org/10.1007/978-981-16-3135-1_27

metal slips in red hot clay mold some of this resistance can be overcome by the high fluidity due to the minimum drop of liquid temperature. The experiments follow the traditional casting technique, and it was so designed to understand the effect of gating system on dimensional accuracy, surface quality, and metallurgical properties on casting thickness. Both open (unrestricted) and restricted pattern dimensions were considered during experimentation.

The investigated data on the cast products of the given properties was then analyzed by a very popular multi-criteria decision-making (MCDM) method—technique for order of preference by similarity to ideal solution (TOPSIS).

2 Background

Indian subcontinent has a very sophisticated non-ferrous investment casting tradition from ancient age (especially Cooperage) [3]. Two important centers were visited to understand the problems of the technology.

i. Dariapur, Guskara, Burdwan District, West Bengal (23°29′1″ N 87°44′6″ E), India.
ii. BiknaShilpaDanga, Bankura District, West Bengal (23°15.3″ N 87°5.9″ E), India.

Most of the utensils and icons produced by them are hollow castings, with 0.5–3.0 mm thick sections. The sections of the castings can be categorized by restricted and open (non-restricted) sections.

The materials used [4] for the wax pattern and clay mold are given in Table 1 with detailed analysis. The steps of the casting technique are summarized below:

(i) Wax pattern (for solid casting)/wax pattern using wax sheets and threads over core (for hollow casting) was made, (ii) several layers of clay over wax pattern were pasted, (iii) drying was done, (iv) gating system (Sprue and cup) was added to exposed wax pattern, (v) the mold was heated inside a coal-fired furnace to red hot condition, (vi) the liquid metal (separately melted) was poured in red hot mold, (vii)

Table 1 Base exchange capacity (sodium equivalent) and grain fineness number of the clay

Material used	Component
Wax pattern	Bees wax—60% and sal-dammar resin 40%
Core clay	Fine kaolinite + sand [Sodium equivalent (miliequivalent/100 g clay) = 3.73] [GFN = 100.68]
Primary coating (Clay mold)	Very fine kaolinite [Sodium equivalent = 0.365] [GFN = 124.97]
Secondary coating (Clay mold)	Kaolinite + rice husk + cow dung + jute cuttings + sand [Sodium equivalent = 1.71] [GFN = 60.7]

Fig. 1 Few defects of "Dokra" casting samples

fettling after cooling was done, and (viii) finishing was conducted. Surfaces near the gate and thicker sections were found poorer to thinner section (Fig. 1).

To improve the quality of investment castings, many researchers tried [5–9] different techniques. The method adopted by Horáček [10] was taken into consideration for assessment of parameters regarding pattern design.

3 Experimental Methodology

Three factors were given importance: (i) Gating system, (ii) casting thickness, and (iii) the presence of core. Quality was optimized by measuring dimensional variation and surface quality.

3.1 Design of Experiment

Three types of gating systems were designed for the experiment—(i) Top gating, (ii) side gating, and (iii) bottom gating. Two types of thickness (w) were chosen for patterns—(i) Thick (w > 3 mm) and (ii) thin (w < 2 mm). The presence of cores is another important factor during pattern dimensions—(i) restricted (cored castings) and (ii) unrestricted/open (solid castings).

The sample numbers according to these factors were given in Table 2. The schematic diagrams of the samples were shown in Fig. 2, which are followed by the proposed model of Horáček [10].

Table 2 Details of the samples

Thin plate (Thickness < 2 mm)			Thick Plate (Thickness > 3 mm)		
S. no.	Type of pattern	Gating system	S. no.	Type of pattern	Gating system
1	Solid	Side	7	Solid	Side
2	Solid	Top	8	Solid	Top
3	Solid	Bottom	9	Solid	Bottom
4	Cored	Side	10	Cored	Side
5	Cored	Top	11	Cored	Top
6	Cored	Bottom	12	Cored	Bottom

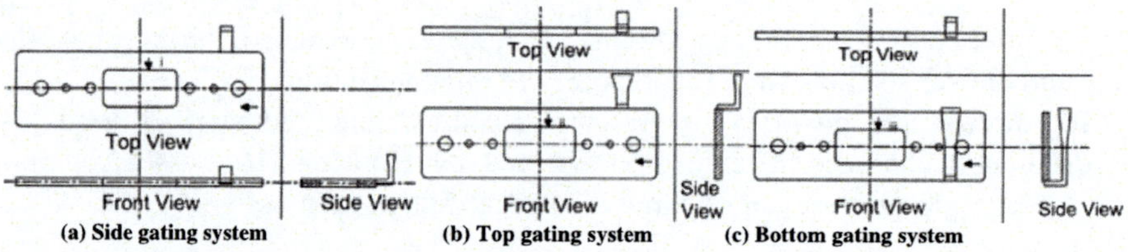

Fig. 2 Schematic diagram of pattern design

3.2 *Experimental Procedure*

A laser cut acrylic thermoplastics dies were designed to produce the wax patterns (Fig. 2). The steps of "Dhokra casting" techniques are shown in Fig. 3. An alloy of Cu–Zn, (60/40) brass was used for casting.

4 Results and Discussions

The cast samples were shown in (Fig. 4) indicating the salvaged and damaged areas.

4.1 *Mechanical Analysis of Cast Samples*

The dimensional variation and surface roughness were measured, following ISO standards ISO 1:2016 [11] and ISO 5436-2:2012 [12], respectively.

- Uniformity of the metal thickness: Root mean square (RMS) technique was used to measure the average thickness variation of all samples (Table 3).

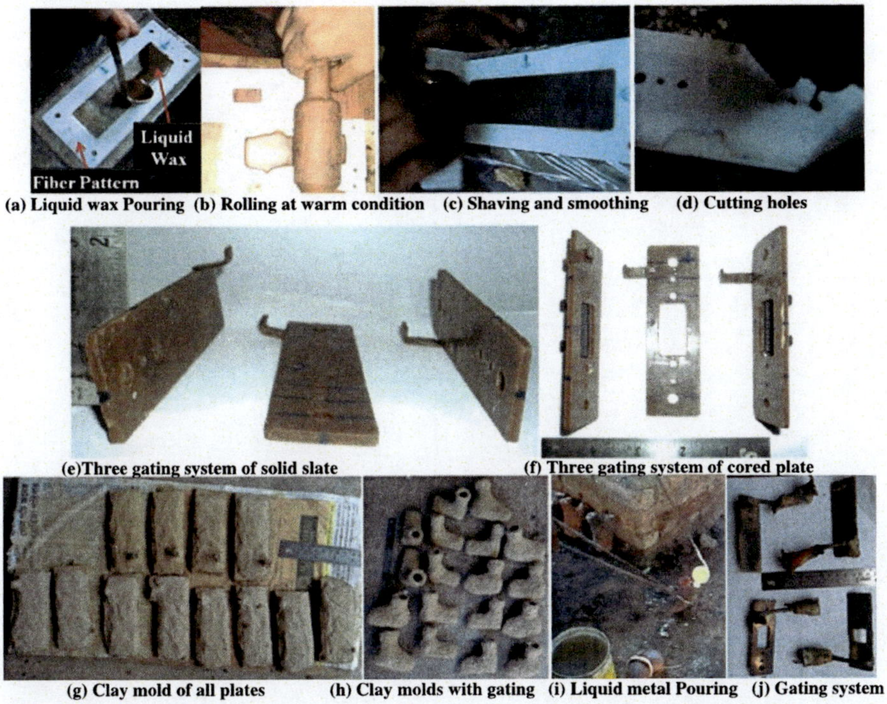

Fig. 3 Procedure of plate making

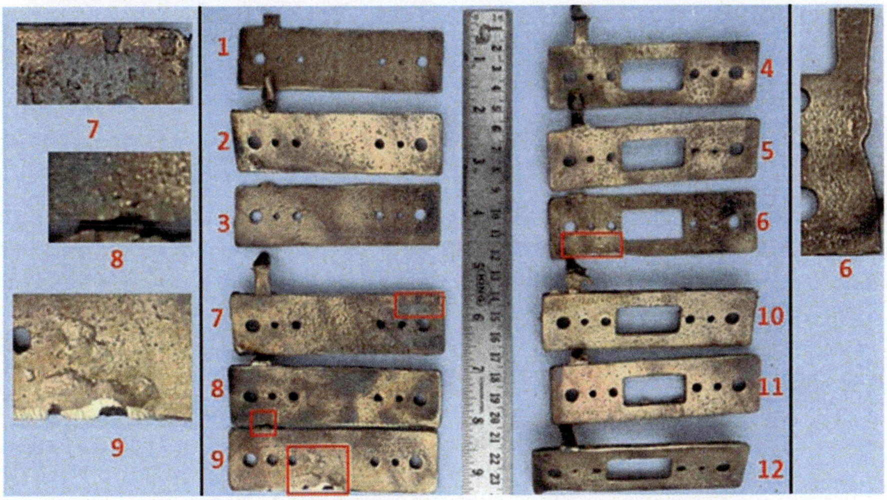

Fig. 4 Surface with enlarged view of casting defects of cast plates with sample no

Table 3 Percentage of mean variation (R.M.S) of the dimensions of the plates

Pattern type	Thin plate ($w < 2$ mm)						Thick plate ($w > 3$ mm)					
	Solid			Cored			Solid			Cored		
Gating type	Top	Side	Bottom	Top	Side	Bottom	Top	Side	Bottom	Top	Side	Bottom
Sample number	1	2	3	4	5	6	7	8	9	10	11	12
Thickness	15.62	18.51	12.7	23.8	14.7	12.74	16.2	11.34	28.2	13.38	9.27	8.23
Longitudinal	0.54	0.4	1.08	1.025	0.925	1.33	1.12	1.895	1.255	1.065	0.965	5.135

- Uniformity of longitudinal dimension (length and breadth): The average of length and breadth, were calculated by the RMS method with computing the discrepancies of the longitudinal dimensions (Table 3).
- Surface quality: The percentage of the salvaged and damaged surface areas were measured by mapping system (Table 4).
- Surface roughness: (Ra): Surface roughness [The arithmetic mean of departures (Ra)] [13] of the samples measured with the sampling length (LC) of 0.8 mm (Fig. 5).

4.2 Characterization

The corrected chemical compositions (wt.%) for zinc equivalent [14] have been assessed and given in Table. 5, along with bulk hardness of the cast samples. The SEM microstructures (Fig. 6) of the cast samples expose the dendrites (which are the signatures of casting process) as well as grain boundaries. The dendrites of the cast metal consist of copper rich α-Cu phase (gray colored) as the major phase ("First-to-freeze" Lean in solute [here Zn]).

4.3 Gradation by TOPSIS

The criteria for gradation were (i) surface quality (less damaged surface is better), (ii) surface roughness (less Ra value is better), (iii) longitudinal dimensional variation (less variation is better), and (iv) metal thickness variation (less variation is better). These criteria are non-commensurable, so, to rank the products, according to acceptance practice, a very popular MCDM method TOPSIS [15–18] was used. To determine the rank of the products, the criteria were chosen in such a way: (i) The lowest values were the best quality; (ii) Equal weight (0.25) was given to determine the rank. The individual and overall ranking of the products were given in Table 6.

4.4 Discussions

The observations of the experiment were detailed below

i. The variation of metal thickness is more in the thin plate than in the thick plate.
ii. The inner mold-wall near the gate expanded and damaged the mold surface for bottom gating system. This happened, due to the slow liquid metal flow (more heat flow to the mold surface) in this case. It causes the surface near the gate poor with high surface roughness and more salvaged and damaged area could be found in sample no's—6, 9, and 12, which were cast using for bottom gating system.

Table 4 Sectional quality and surface roughness of the plates

Gating type		S. no	Thin plate ($w < 2$ mm)			S. no	Thick plate ($w > 3$ mm)		
			Surface quality (%)		Roughness (Ra)		Surface quality (%)		Roughness (Ra)
			Salvaged	Damaged			Salvaged	Damaged	
Unrestricted (solid)	Side	1	0	0	1.415	7	0	0	0.0112
	Top	2	0	0	1.29	8	0.07	0	0.01895
	Bottom	3	1.55	0	2.275	9	2.57	0	0.01255
Restricted (cored)	Side	4	0	0	4.265	10	2.55	0	0.01065
	Top	5	0	0	4.92	11	0.01	0	0.00965
	Bottom	6	2.81	2.24	3.185	12	2.92	2.24	0.05135

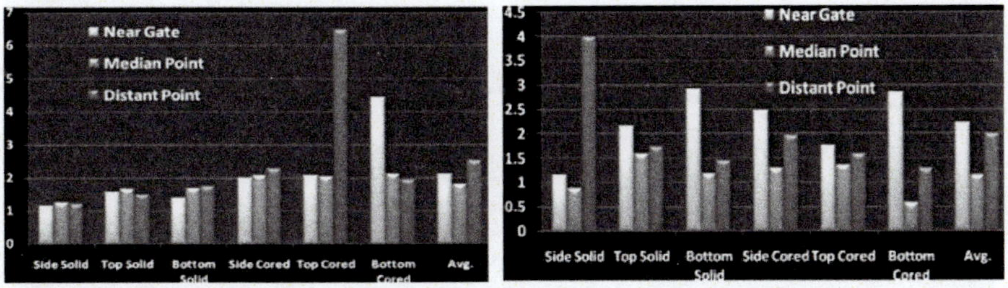

Thin Plate (w<2) **Thick Plate (w>3)**

Fig. 5 Surface roughness (Ra) with respect to distance from gate

Table 5 Chemical composition and bulk hardness

Chemical composition (wt%)					Bulk hardness (HV 5/10)	
Cu	Zn	Sn	Pb	Fe	Thin plate	88.0
62.5	30.39	1.39	4.18	1.04	Thick plate	84.6

Zn Equivalent: 36.18, **Cu–Zn Ratio**: 63.82: 36.18

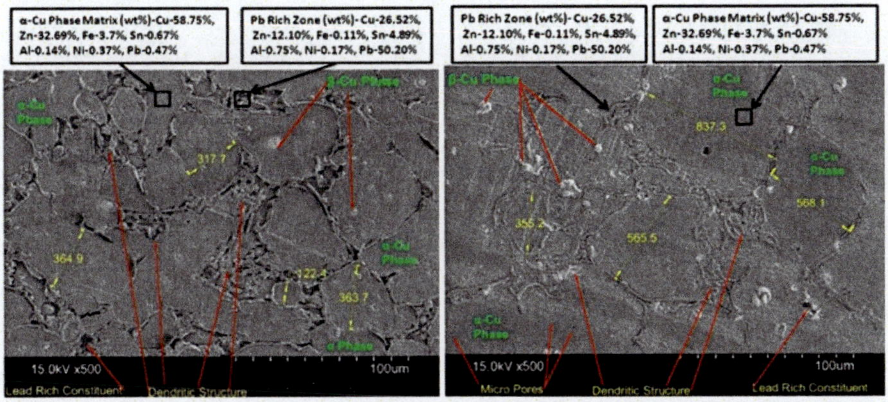

Fig. 6 Microstructure (SEM) of thin plate (top) and thick plate (bottom) (500×) with EDX analysis (Etchant: $FeCl_3$ in HCl). The structure shows coarse **α**-Cu phase as matrix with inter dendritic region and also with minor amount of **β**-Cu phase (white patches). Insoluble Pb is present at few points. Dimension of grains shows that thin casting contains smaller grains compare to thinner one (All dimensions are in μm)

iii. The smooth liquid metal flow was obstructed due to presence of cores, especially for thin casting. As a result, the surface roughness of (Ra) thin plates was better for unrestricted (solid) castings compared to restricted (cored) ones. But for thick castings, the core has a negligible effect. Because in this case the liquid-to-solid metal shrinkage property only controls the surface quality.

Table 6 Ranking of the casting samples based on quality using the TOPSIS method

Ranking	Thin plate (thickness < 2 mm)						Thick plate (thickness > 3 mm)					
	Unrestricted			Restricted			Unrestricted			Restricted		
	Top	Side	Bottom	Top	Side	Bottom	Top	Side	Bottom	Top	Side	Bottom
Sample no	1	2	3	4	5	6	7	8	9	10	11	12
Individual	1	2	5	3	4	6	1	2	4	3	5	6
Overall	1	2	5	6	7	11	3	4	9	8	10	12

iv. The surfaces distant from the gate had more shrinkage defect [19]. Poor rising may be the cause of unacceptable surface for sample no's 5 and 7.

v. In microstructures, coarse (large) grains were present in thick plates, whereas (small) fine grains are present in the thin plates. Quite a large number of dendrites are present in thin plates as faster cooling happened, which gives rise to large number of nucleation sites without consequent available time for growth. Smaller the thickness, lower is the solidification time, as per 'Chvorinov Rule,' the solidification time is directly influenced by of the ruling section, following, [20]

$$w = k\sqrt{t} - c \quad (1)$$

where $w =$ solidification thickness, $t =$ solidification time, and k, c are constant. Therefore, the experimentation had shown that the grain size of a cast sample depends on the casting thickness.

iv. Side gating and top gating systems produced castings having better acceptable surface compared to bottom gating system. So, bottom gating system is not followed for thin casting production.

5 Conclusion

On the basis of the investigation, the following conclusions can be drawn.

i. In the field visit, the major defects of the castings, identified in thin sections, were misrun and rat tails. In thick sections, the main defects found were crushes, metal penetration, scab, and pits.

ii. Design of gating system is important for good quality casting and economic production. The experiment proves that gates should not be positioned at the corner of any sample; rather this should be positioned in such a way that every corner should be filled quickly [21] from the principle of gating design: "Shortest filling time."

iii. Restricted cast components (castings having core) and thin sections require more care and accuracy for better quality and economic production. The castings having thick sections require proper riser design to avoid shrinkage defects.

iv. The grains structure and size are controlled by the ruling thickness of a component. So, the heat treatment and machining of a component can be reduced by proper design of a component and production method selection.

Acknowledgements This work was funded by AICTE (All India Council for Technical Education) through AICTE Doctoral Fellowship (ADF). The author is thankful to the Department of Metallurgical and Material Engineering of Jadavpur University and Mr. Gopan Karmakar—Artisan of Bikna, for their cooperation.

References

1. Roy S et al (2019) Comparative study of dimensional variation and surface characteristics of thin walled investment casting of different geometrical shape for different gating system in hot clay mold. IOSR J Eng 09(3):17–23. http://iosrjen.org/Papers/vol9_issue3/Series-2/B0903021723.pdf
2. Roy S et al (2017) Kinetics of liquid metal flow in gating design of investment casting production. SLÉVÁRENSTVÍ (Association of Foundries of the Czech Republic) vol 5–6:149–154. https://issuu.com/inasport/docs/slevarenstvi_5_6_2017/18
3. Mandal B, Datta PK (2010) Hot mold casting process of ancient east India and Bangladesh. China Foundry 171–177:171–07. 1672–6421(2010)02
4. Roy S et al (2019) The technology inherent in Indian traditional craft metal casting process. Indian Foundry J 65(8):27–37
5. Donga YW et al (2017) Modelling of shrinkage during investment casting of thin-walled hollow turbine blades. J Mater Process Technol 190–203. https://doi.org/10.1016/j.jmatprotec.2017.01.005
6. Cannell N, Sabau AS (2005) Predicting pattern tooling and casting, dimensions for investment casting, phase II. U.S. Department of Energy (DOE) Information Bridge. https://doi.org/10.2172/974578
7. Faguo L et al (2015) Mechanism of filling and feeding of thin-walled structures during gravity casting. Materials 3701–3713. https://doi.org/10.3390/ma8063701
8. Sabau AS (2006) Alloy shrinkage factors for the investment casting process. Metall Mater Trans B 131–114. https://doi.org/10.1007/s11663-006-0092-x
9. Raza M (2015) Experimental study of the filling of thin-walled investment castings In 17–4ph stainless steel. Metall Foundry Eng 41(2):85–98. https://doi.org/10.7494/mafe.2015.41.2.85
10. Horáček M (2005) Accuracy of investment casting. Arch Foundry 5(15):121–137. https://pdfs.semanticscholar.org/9e55/9c68eff67f7ee59bfda2bc9a5e127da2e572.pdf
11. ISO 1: 2016: Geometrical product specifications (GPS)—standard reference temperature for the specification of geometrical and dimensional properties. https://www.iso.org/standard/67630.html
12. Geometrical product specifications (GPS)—surface texture: Profile method; Measurement standards—part 2: software measurement standards. https://www.iso.org/standard/61261.html
13. Dotson CL (2005) Fundamentals of dimensional metrology, 375, Cengage Learning
14. Mandal B, Datta PK (2010) Understanding alloy design principles and cast metal technology in hot molds for medieval Bengal. Indian J History Sc 101–140. https://insa.nic.in/writereaddata/UpLoadedFiles/IJHS/Vol45_1_5_BMandalarticle.pdf
15. Mardani A et al (2015) Multiple criteria decision-making techniques and their applications—a review of the literature from 2000 to 2014. Econ Res-Ekonomska Istraživanja 28(1):516–571. https://doi.org/10.1080/1331677X.2015.1075139
16. Renato AK, André GC (2015) Pacheco: a-TOPSIS—an approach based on TOPSIS for ranking evolutionary algorithms. Proc Comput Sci 55:308–317. Elsevier. https://doi.org/10.1016/j.procs.2015.07.054
17. Silva DFL, Filho ATA (2020) Sorting with TOPSIS through boundary and characteristic profiles. Comput Industr Eng 141:308–317. Elsevier. https://doi.org/10.1016/j.procs.2015.07.054
18. Raju KS, Kumar DN (2010) Multicriterion analysis in engineering and management, pp 80–81. PHI Learning Pvt. Ltd.
19. Campbell J (2015) Complete casting handbook, 2nd ed, pp 466–470. Elsevier. https://doi.org/10.1016/C2014-0-01548-1
20. Sylvia JG (1972) Cast metal technology, p 140. Addison-Wesley Publishing Co.
21. Roy S et al (2020) Precise filling time calculation of thin-walled investment casting in hot mold (In press). J Braz Soc Mech Sci Eng. https://doi.org/10.1007/s40430-020-02634-6

Tool Wears in Milling of Al/SiCp Composites by Dry and MQL Milling

Ankush Kohli, H. S. Bains, and Sumit Jain

Abstract The milling of aluminium silicon carbide reinforcement metal matrix composites (Al/SiCp MMCs) under two machining environments (one is dry and second is MQL) are analysed. The objective of this work is to reduce the flank wear during milling of Al/SiCp MMCs by the application of MQL. The paper not working only to reduce the tool wear but also worked on milling input parameters like feed rate, speed and depth of cut on tool wear. The forecast of the optimum value of the tool wear during milling of Al/SiCpMMCs plays a crucial role in analysing the working surface and tool life through the latest optimization techniques. The forecast model represented the impact of milling responses on tool wear (V_{bmax}). Then, the forecast-model result was compared with the experimental result of milling of Al/SiCp MMCs and observed to be almost same with them. The forecast and experimental results show that predicted model helps in to set the milling input parameters to minimizing flank wear V_{bmax}, during milling of Al/SiCp MMCs. Also, the results showed that application of minimum quantity lubrication helps in reduction of flank wear during milling of Al/SiCp MMCs.

Keywords Milling · Al/SiCp MMCs · Tool flank wear · MQL

1 Introduction

Aluminium silicon carbide metal matrix composites (Al/SiCp MMCs) have been introduced in many engineering applications for more than twenty years because of their superior mechanical properties like hardness, strength and wear resistance [1]. The substitution of customary material shows that this mixture of reinforcement

A. Kohli (✉)
Mechanical Engineering, IKGPTU, Punjab, Jalandhar, India

H. S. Bains
Department of Mechanical Engineering, PUSSGIRI, Punjab, Hoshiarpur, India

S. Jain
Department of Mechanical Engineering, CTIEMT, Punjab, Jalandhar, India

A. K. Dubey et al. (eds.), *Recent Trends in Industrial and Production Engineering*, Lecture Notes in Mechanical Engineering,
https://doi.org/10.1007/978-981-16-3135-1_28

of SiCp in aluminium matrix offers a decent potential in the automotive and aviation industry [2]. These materials are additionally ready to be utilized in the sports industry, aluminium pipe vessel and also developing wind energy from wind turbine edges, and even also seen in gas sewerage pipe [3, 4]. The capability of the numerical control milling machine to manufacture any fixed or any complex shape is another favourable advantages for Al/SiCp MMCs.

The troubles of the milling of Al/SiCp MMCs are one of the serious issues that forecast its broad use in designing applications [5–7]. These Al/SiCp MMCs materials are anisotropic, non-homogeneous and the hard nature of their enforcement (SiCp) are exceptionally grating, which makes these materials hard to machine. Subsequently, the utilization of Al/SiCp MMCs in the automotive and aviation industry has been restricted because of the higher machining costs and shorten tool life because of high tool wear rate during milling of Al/SiCp MMCs. However, by using a smaller grain size of the reinforcement in the fabrication of Aluminium MMCs, their mechanical properties as well as machining properties can be enhanced. At the same time, the researcher concentrates on the milling parameter of Al/SiCp MMCs to improve the life of milling inserts but not compromise with the quality of the operational surface of the composites. Sahin et al. evaluated the tool wear for aluminium MMCs for two different tools and concluded that TiN tool has marvellous life as compared to the TP30 tool and also mentioned that tool life decreased for both the tool with increase in the speed [8]. Jeyakumar et al. predicted the tool wear of only dry milling of Al/SiCp MMCs using RSM and stated that more tool wear rate was observed at low cutting speed and also formation of BUE at tool face [9]. So the literature work predicted that there is some gap between the milling of AlSiCp MMCs by dry and MQL conditions.

2 Materials for Experimentation

Aluminium (Al) block of dimension 115 × 35 × 50 mm for AL6061/Sicp MMC is fabricated through stir casting in this experimentation. In this, Al is pre-heated to a low temperature of above 500 °C and then further heated in the furnace up to 950 °C. After that, mixing of hard SiCp reinforcement in the molten Aluminium and a vortex is produced through a stirrer having a speed of around 475 rpm for at least 35 minutes and then pour molten Al/SiCp in the die. The Al-block having a size of 115 × 34 × 48 mm is obtained by providing a roughing cut through flexible CNC vertical milling machine for the milling of Al/SiCp composites under dry and minimum quantity lubrication (mustard oil) conditions. The carious composition of Al/Sic_p MMCs is as shown in Table 1.

For each cut in the surface, different carbide face of the insert is used and also performed this cut on a fresh surface of MMCs having a fixed length of 115 mm is taken into consideration. Digital Mitutoyo is employed for measuring flank wear to obtain the accurate tool wear value. For the milling of Al/SiCp composites, a triangular-shaped insert is used with a nose radius of the insert is 0.4 mm as depicted

Table 1 Various elements of Al6061/Sic$_p$ MMCs

Elements	Si	Fe	Cu	Mn	Mg	Cr
Al/SiC$_p$MMC	0.251	0.29	0.032	0.03	0.25	0.018

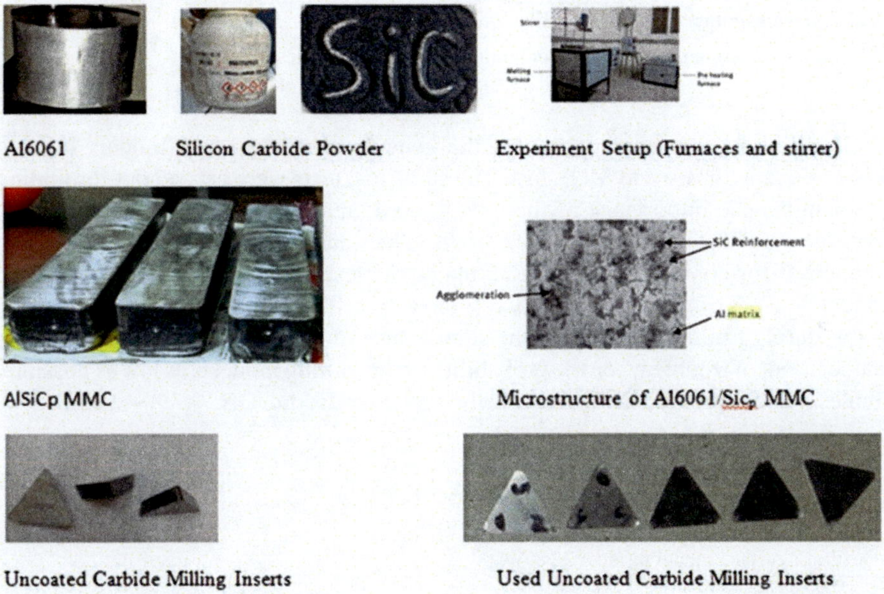

Fig. 1 Material Al and Al/SiCp MMCs, reinforcement SiCp, experiment setup and inserts

in the Fig. 1. The complete information for this experiment, that is, Al-6061, silicon-carbide reinforcement, a stirrer, furnace and Al/SiCp MMCs block, Al/SiCp MMcs microstructure and carbide inserts are shown in Fig. 1.

3 Experimental Design

The milling design with carbide inserts involved with different sets of milling input variable under two machining environments, i.e. dry and MQL conditions, is depicted in the table. Box-Behnken design is used for this experimentation. The purpose of the selection of this design is simple, least cost, the low number of cut and consume less time as compared with the CCD design. No axial points are used in this design so that all the milling input and output variables keep in the safe zone and hence keep it safe. A three layer of this milling design in the RSM approach was performed to obtain the optimum milling response for finding tool wear V_{bmax} of Al/SiCp MMCs (Table 2).

Table 2 Milling machine and their responses

Milling	Milling HUNCO VM10
Cutting-inserts	Carbide milling insert by Widea
Cutting response	Speed S: 2000, 3000 and 4000 rpm, feed rate F: 200 to 400 mm/min. and doc: 1.0 to2.0 mm
Tool wear apparatus	Digital Mitutoyo
Machining environments	Dry (D) and minimum quantity lubrication (M1)

The designs test was applied for the generation of the linear models for tool wear ($V_{bmax.}$) for dry and MQL test. The statistical tests were carried out for finding the significance importance of the model; no doubt, it also depends on individual variables and determining the lack of fit. The analysis of variance (ANOVA) is connected for condensing and recapitulates the tests performed in this study. The model f-values of the tool wear ($V_{bmax.}$) model are 12.852 for both dry as well as for MQL during milling represented the significance of the milling model. The milling model is out of trouble when the probability of the milling model is as low as possible. Tables 3 and 4 provide the results for the tool wear from DOE.

Table 3 Milling model for tool wear V_{bmax} .for dry milling

Parameter-1	$V_{bmax.}$ (D)					
Quadratic surface model through ANOVA						
Variance table						
Response	SSQ	Df	MSq	Value of F	Prob > F	
Milling model	0.279	9	0.031	12.852	0.0014	Significant
A-Sr	0.185	1	0.185	76.749	<0.0001	
B-F	0.001	1	0.001	0.284	0.6104	
C-Doc	0.001	1	0.001	0.540	0.4863	
AB	0.008	1	0.008	3.439	0.1061	
AC	0.004	1	0.004	1.545	0.2539	
BC	0.052	1	0.052	21.775	0.0023	
A^2	0.001	1	0.001	0.401	0.5466	
B^2	0.008	1	0.008	3.180	0.1177	
C^2	0.017	1	0.017	6.862	0.0344	
Residual	0.017	7	0.002			
Lack of fit	0.013	3	0.004	4.979	0.0775	Not significant
Pureerror	0.004	4	0.001			
Cor total	0.295	16				

Table 4 Milling model for tool wear V_{bmax} .for MQl milling

Parameter-2	$V_{bmax.}$ (M)					
Quadratic surface model through ANOVA						
Variance table						
Response	SSQ	Df	MSq	Value of F	Prob > F	
Milling model	0.268	9	0.030	12.852	0.0014	Significant
A-Sx	0.178	1	0.178	76.749	<0.0001	
B-F	0.001	1	0.001	0.284	0.6104	
C-Doc	0.001	1	0.001	0.540	0.4863	
AB	0.008	1	0.008	3.439	0.1061	
AC	0.004	1	0.004	1.545	0.2539	
BC	0.050	1	0.050	21.775	0.0023	
A^2	0.001	1	0.001	0.401	0.5466	
B^2	0.007	1	0.007	3.180	0.1177	
C^2	0.016	1	0.016	6.862	0.0344	
Residual	0.016	7	0.002			
Lack of fit	0.013	3	0.004	4.979	0.0775	Not significant
Pureerror	0.003	4	0.001			
Cor total	0.284	16				

4 Modelling of Tool Wear ($V_{bmax.}$)

The parametric representation of tool wear of milling of Al/SiCp MMCs under dry and MQL machining environment has been generated with milling responses as shown in the Eqs. 1 and 2.

$$V_{b\max.}(Dry) = 0.45 + 0.15 * A + 0.0093 * B + 0.013 * C - 0.046 * A * B - 0.031\, A * C + 0.11 * B * C - 0.015 * A^2 - 0.043 * B^2 - 0.063 * C^2. \quad (1)$$

$$V_{b\max.}(MQL) = 0.44 + 0.15 * A + 0.0090 * B + 0.012 * C - 0.045 * A * B - 0.03\, A * C + 0.11 * B * C - 0.015 * A^2 - 0.042 * B^2 - 0.061 * C^2. \quad (2)$$

The milling model forecasts the values of tool wear $V_{bmax.}$ of milling of Al/SiCp MMCs on other milling variables. The current mathematical model is adopted for forecasting of the flank wear of milling of Al/SiCp MMCs.

5 Analysis of Results and Optimization

The result for flank wear of milling of Al/SiCp MMCs under two milling condition (dry, D and MQL, M1) are depicted in Table 5. The results showed that flank wears increase with an increase in the Doc and also with feed rate.

The minimum value is of flank wear is 0.253 mm in dry machining condition at a speed above 2000 rpm and less than 2250 rpm, feed rate: 200 to 260 mm/min. at a constant depth of cut, 1.5 mm corresponding to the minimum value of flank wear 0.248 mm in minimum quantity lubrication (MQL) at same machining input variables. This value of flank wear in second machining environments is lesser than dry machining environment which clears that application of MQL helps in the reduction of flank wear as appeared in the Fig. 2.

The graphical representation of results for milling of Al/SiCp MMCs for Box-Behnken design under two machining environment is shown in the Fig. 3. The results show that second milling operation, that is, by the application of MQL is better than dry milling.

The optimum value of the flank wear is 0.18 mm and 0.17 mm at speed: 2567.92 rpm, feed rate: 395.64 mm/min. and depth of cut: 1.01 mm for dry (D) and minimum quantity lubrication (M), respectively, according to the DOE as shown in Table 6.

Table 5 Flank wear of milling of Al/SiCp MMCs for two machining environments

S. No.	Speed (rpm)	Feed (mm/min.)	Doc (mm)	V_{bmax} (Dry)	V_{bmax} (M1)
1	4000	400	1.5	0.51	0.50
2	2000	300	1	0.18	0.18
3	3000	200	2	0.19	0.18
4	4000	300	2	0.50	0.49
5	2000	400	1.5	0.25	0.24
6	3000	400	1	0.27	0.27
7	2000	300	2	0.31	0.30
8	4000	300	1	0.50	0.49
9	3000	200	1	0.44	0.43
10	3000	300	1.5	0.48	0.47
11	3000	300	1.5	0.40	0.39
12	3000	300	1.5	0.46	0.45
13	3000	300	1.5	0.46	0.45
14	4000	200	1.5	0.63	0.62
15	3000	300	1.5	0.45	0.44
16	2000	200	1.5	0.18	0.18
17	3000	400	2	0.48	0.47

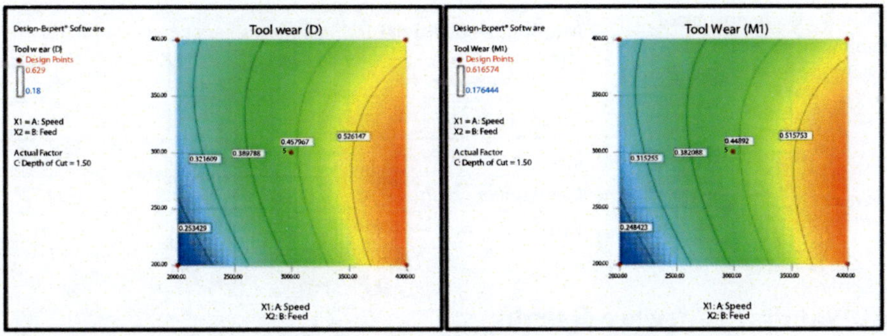

Fig. 2 Milling response for tool wear under dry and MQL environments

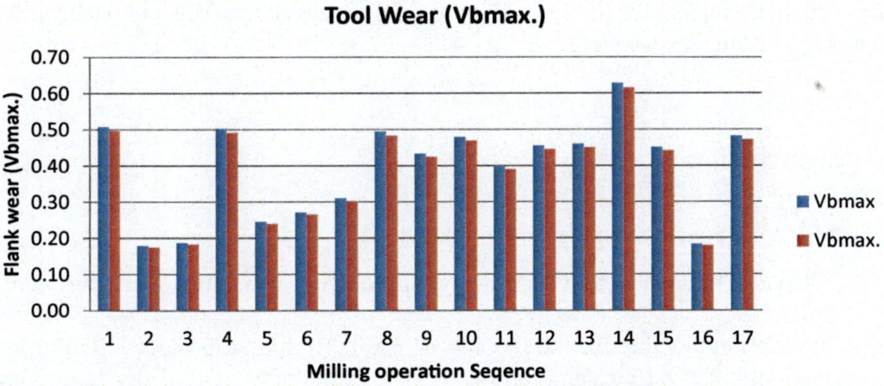

Fig. 3 Flank wear for milling of Al/SiCp MMCs under dry and MQL environments

Table 6 Solutions for flank wear for two machining environments from DOE

Number	Speed	Feed	Depth of cut	Tool wear (D)	Flank wear (M1)	Desirability	
1	2567.92	395.64	1.01	0.18	0.17	1.00	Selected
2	2008.34	397.98	1.11	0.14	0.14	1.00	
3	2017.95	226.14	1.89	0.15	0.15	1.00	
4	2205.34	383.93	1.02	0.14	0.14	1.00	
5	2298.15	208.42	1.89	0.17	0.16	1.00	
6	2031.40	200.15	1.64	0.17	0.17	1.00	
7	2118.00	201.48	1.69	0.18	0.17	1.00	
8	2028.15	352.59	1.06	0.17	0.16	1.00	
9	2027.70	243.96	1.99	0.16	0.16	1.00	
10	2062.93	358.51	1.04	0.16	0.16	1.00	

Table 7 Comparisons of flank wear results under two machining environments

Machining condition (dry and M1)	Process parameters (in range) S_x: 2567, F_r: 395 mm/min. and DoC: 1.01 mm	
Values of V_{bmax}	Dry	MQL
Software result	0.18	0.17
Experiment results	0.183	0.172
Error	0.02	0.01

6 Validate Software Results

It was seen that the experimental result is same as software results at optimum milling conditions as depicted in Table 7. The errors in the result are 0.02 and 0.01 for dry and MQl results, respectively.

7 Conclusion

The following conclusion may be adopted

1. The value of flank wear is high at slow speed due to generating high operational forces.
2. The optimum value for the flank wear was 0.18 mm at the speed: 2567 rpm, feed rate: 395.64 mm/min. and DOC: 1.01 mm as in case of dry machining condition.
3. The optimum value for the flank wear was 0.17 mm at the speed: 2567 rpm, feed rate: 395.64 mm/min. and DOC: 1.01 mm as in case of MQL machining condition (M1).
4. The experimental results validated the software results; hence, current model for predicting the flank wear of Al/SiCp MMCs is valid.

Acknowledgements We thank the IKGPTU, KAPURTHALA for providing the platform to encourage the research in the area of machining of composites. I am also thanks to the current conference proceeding for published my research work in the area of machining of composites by different environments.

References

1. Ramanujam R, Venkatesan K, Kothawade N, Shivangkumar J (2015) Fabrication of Al-TiB 2 metal matrix composites for evaluation of surface characterization and machinability. Indian J Sci Technol 8:85–89

2. Zalnezhad E, Sarhan AAD, Hamdi M (2013) Optimizing the PVD TiN thin film coating's parameters on aerospace AL7075-T6 alloy for higher coating hardness and adhesion with better tribological properties of the coating surface. Int J Adv Manuf Technol 64(1–4):281–290
3. . Ravi Shankar M, Chandrasekar S, Compton WD, King AH (2005) Characteristics of aluminum 6061-T6 deformed to large plastic strains by machining. Mater Sci Eng A 410–411:364–368
4. Lakshminarayanan AK, Balasubramanian V, Elangovan K (2009) Effect of welding processes on tensile properties of AA6061 aluminium alloy joints. Int J Adv Manuf Technol 40(3–4):286–296
5. Krishnaraj V, Raghavendran N, Sudhan R, Vignesh R (2012) An investigation on end milling of aluminium based metal matrix composites for optimising machining parameters. JMFT 4
6. Sharma VS, Singh G, Sørby K (2015) A review on minimum quantity lubrication for machining processes machining processes, vol 6914, no. February 2016
7. Barnes S, Pashby IR, Group WM (1995) Machining of aluminium based metal matrix composites. Appl. Compos Mater 31–42
8. Sahin Y, Kok M, Celik H (2002) Tool wear and surface roughness of Al_2O_3 particle-reinforced aluminium alloy composites. J Mater Process Technol 128:280–291
9. Jeyakumar S, Marimuthu K, Ramachandran T (2013) Prediction of cutting force, tool wear and surface roughness of Al6061/SiC composite for end milling operations using RSM. J Mech Sci Technol 27(9):2813–2822